Bibliographic information published by the Deutsche Nationalbibliothek

The Deutsche Nationalbibliothek lists this publication in the Deutsche Nationalbibliografie; detailed bibliographic data are available in the Internet at http://dnb.d-nb.de .

ISBN 978-3-8325-2962-8

Logos Verlag Berlin GmbH
Comeniushof, Gubener Str. 47,
10243 Berlin
Tel.: +49 (0)30 42 85 10 90
Fax: +49 (0)30 42 85 10 92
INTERNET: http://www.logos-verlag.de

Systematic characterization of HT PEMFCs Containing PBI/H$_3$PO$_4$ systems

Von der Fakultät für Ingenieurwissenschaften,

Abteilung Maschinenbau und Verfahrenstechnik

der

Universität Duisburg-Essen

zur Erlangung des akademischen Grades

eines

Doktors der Ingenieurwissenschaften

Dr.-Ing.

genehmigte Dissertation

von

George Chakravarthy Bandlamudi

aus

Tenali, Indien

Gutachter: Univ.-Prof. Dr. rer. nat. Angelika Heinzel

PD. Dr. rer. nat. Werner Lehnert

Tag der mündlichen Prüfung: 01.Juli 2011

Acknowledgements

I would like to express my sincere gratitude to my thesis adviser and guide, Prof. Dr. rer. nat. Angelika Heinzel, Chair of the Institute for Energy and Environmental Protection Technologies, Department of Mechanical Engineering, at the University of Duisburg-Essen, Campus Duisburg, for her constant support and very useful discussions in the course of this work. Experimental investigations were performed at the centre for fuel cell technology (Zentrum für Brennstoffzellentechnik or ZBT) in Duisburg, Germany. I am thankful to Dr. rer. nat. Falko Mahlendorf for his inputs. I am grateful to Dr. Ing. Jürgen Roes for his useful hints and inputs.

I am very grateful to PD. Dr. rer. nat. Werner Lehnert from the German National Research Centre (Forshungszentrum), Jülich, Germany for his very constructive and useful inputs and for being the second supervisor for this work.

I am also grateful to my friends and colleagues at ZBT GmbH, Duisburg and also my friends and colleagues at the Institute for Energy and Environmental Protection Technologies, Department of Mechanical Engineering, University of Duisburg for their constant support.

I am grateful to Huldah my wife and the rest of my family for their constant support and extraordinary patience.

Systematic characterization of HT PEMFCs

Containing PBI/H$_3$PO$_4$ systems

Thermodynamic analysis and Experimental investigations

Abstract

The aim of this work is to identify performance limiting mechanisms and processes that pertain to high temperature polymer electrolyte fuel cells (HT PEMFCs) containing H_3PO_4/PBI based systems, while operating them for long periods of time. In this work, the performance limiting losses such as ohmic, concentration, kinetic and fuel crossover overvoltages are studied, after constructing a HT PEMFC single cell and operating the same in a test stand consisting of all the necessary hardware to characterize the same.

The Ohmic overvoltage (both electronic and ionic) contributions from all the cell components have been studied systematically using electrochemical impedance spectroscope (EIS) analyser. Membrane resistance as seen by the cell is expressed in Arrhenius form, where the pre-exponential factor (σ_0) was found to be 13466 and the activation energy (Ea) was found to be 18484 J/mol.

The anhydrous (conductivity not critically dependant on water) behaviour of highly doped as well as lowly doped membranes relevant to HT PEMFCs has been examined with EIS analyser and the subsequent spectral signatures are documented.

The kinetic parameters governing cell's activation losses such as exchange current density (i_0) and transfer co-efficient (α) have been analysed at different cell operating temperatures. The (i_0) values were found to be 0.83 mA/cm² at 180°C and 0.04 mA/cm² at 130°C. The (α) value was 0.42 at 180°C and was 0.50 at 130°C.

The HT PEMFC's performance at load currents of (0-400 mA/cm²) and temperatures of (130°C-180°C) has been studied. Operating the cell at lower temperatures (~ 130°C) might be interesting during the start up period, as these HT PEMFCs need to be heated up from room temperature to its operating temperature of ~ 160°C. Operating the cell at 130°C is not attractive from the point of view of cell performance and tolerance to fuel impurities (such as CO in reformates). Whereas operating it close to 180°C may not be the best from the point of view of electrolyte retention, catalyst and catalyst support (carbon) stability. Thus, an operating cell temperature of 160°C might be a viable compromise between cell performance and tolerance to impurities, but offering higher durability. Carbon monoxide tolerance of HT PEMFCs at different temperatures and load currents has been studied and optimum conditions for re-former-HT PEMFC coupled systems have been identified.

It was observed that at 180°C, a HT PEMFC fit with a Celtec P® 2000 MEA could of-fer a CO tolerance of up to 20% with a loss of about 150 mV (at 200 mA/cm² of load), whereas the same could not even stand 1% CO at 130°C (the cell had lost almost the entire cell voltage when fed with 1% CO into its anode stream of H_2 at 130°C). A HT PEMFC stack's tolerance to other synthetic reformates (consisting of varying concentrations of hydrogen) has been studied.

HT PEM fuel cell degradation caused due to the loss of phosphoric acid (electrolyte), electrochemically active surface area (ECSA) has been studied.

ICP-OES and ICP-MS techniques have been used to analyse cell's product water, whereas Zahnermesstechnik's IM6 analyser was used to study the ECSA by perform-ing CV (cyclicvoltammetry) scans.

Based on a 1000 hour test, it was observed that a 10°C rise in temperature (from 160°C) resulted in an electrolyte loss rate of > 2.0 times its value at 160°C. The elec-trolyte loss rate was around 0.45 µg/m²/s in the case of 170°C cell, whereas its was around 0.20 µg/m²/s in the case of 160°C cell.

Long term performance (2000 hour cell operation) and accelerated cell degradation (when operated with zero load current or OCV for 120 hours) have been studied.

Cell components' stability has been examined by using SEM (scanning electron mi-croscopy) images.

After the OCV (120 hour) test, it was observed that the loss in cell's ECSA was 20% from its initial value; its voltage loss was 16.6% from its initial value.

From the 2000 hour test, it was observed that the cell had lost 42.2 mV (0.6610 – 0.6188 V)(or voltage loss rate of 6.4%) at an average load of 200 mA/cm², at an av-erage cell temperature of 160°C. That translates to a performance loss rate of 10 µV/hr on constant load (200 mA/cm²) and an average of 20 µV/hr (with constant load, load cycling and temperature cycling, shutdown and start-up phases included).

Zusammenfassung

Das Ziel dieser Arbeit war es, die leistungsbeschränkenden Mechanismen und Prozesse zu identifizieren, die während des Langzeitbetriebes auf die Hochtemperatur-Polymer-Elektrolyt-Brennstoffzellen (HT PEMFC) mit H_3PO_4/PBI basierten Elektrolyt-Systemen einwirken. Zu diesem Zweck wurden in dieser Arbeit die Leistungsverluste erforscht, zu denen sowohl die ohmschen Verluste als auch die Verluste zählen, die aufgrund des Massentransports, der Kinetik und der Brennstoff-Crossover-Überspannungen entstehen.

Eine einzelne HT PEMFC Zelle wurde zu diesem Zweck konstruiert und auf demselben hierfür konfigurierten Teststand betrieben, der aufgrund der gewählten Ausstattung eine Vergleichbarkeit der Messergebnisse gewährleistet. Der ohmsche Überspannungsanteil (elektronische und ionische) aus allen Zellbestandteilen wurde systematisch mit Hilfe eines elektrochemischen Impedanzspektroskopie (EIS)-Analysators untersucht. Der Membranwiderstand, der entlang der Zelle wirkt, entspricht der Arrhenius Form, wobei der präexponentiellen Faktor (σ_0) mit einem Wert von 13466 und die Aktivierungsenergie (E_a) mit 18484 J/mol bestimmt worden sind.

Das maßgeblich anhydrische Verhalten (Leitfähigkeit hängt nicht vom Wassergehalt ab) von sowohl hoch als auch niedrig dotierten HT PEMFC Membranen wurde mit Hilfe des EIS-Analysators und der anschließenden Auswertung der spektralen Signaturen nachgewiesen und dokumentiert.

Die kinetischen Parameter, die die Zellaktivierungsverluste betreffen, wie die Austauschstromdichte (i_0) und des Transfer-Koeffizienten (α) wurden bei verschiedenen Zell-Betriebstemperaturen analysiert. Die (i_0) Werte wurden mit 0,83 mA/cm² bei 180°C und mit 0,04 mA/cm² bei 130°C bestimmt. Der (α)-Wert betrug 0,42 bei 180°C und 0,50 bei 130°C.

Die HT PEMFC Leistung wurde bei Lastströme (0-400 mA/cm²) und Temperaturen von (130°C-180°C) untersucht. Der Betrieb der Zelle bei niedrigeren Temperaturen (~ 130°C) spielt während der Startphase eine Rolle, da die HT PEMFC von Raumtemperatur erwärmt werden muss, um seine Betriebstemperatur von ~ 160°C zu erreichen. Der Betrieb der Zelle bei 130°C weist im Hinblick auf die Zellleistung und die Toleranz gegenüber Verunreinigungen (z. B. CO in Reformat) des Brennstoffes deutliche Schwächen auf. Bei einer Betriebstemperatur von 180°C treten nachteile bezüglich der Elektrolyt Rückhaltung, der Katalysator und Katalysatorträger-(Kohlenstoff) Stabilität auf. So könnte eine Zell-Betriebstemperatur von 160°C ein tragfähiger Kompromiss zwischen Leistung der Zelle und der Toleranz gegenüber Verunreinigungen sein, der eine höhere Standzeit erlaubt.

Die Kohlenmonoxid-Toleranz von HT PEMFC bei unterschiedlichen Temperaturen und Lastströmen wurde untersucht und optimale Bedingungen für einen gekoppelten Betrieb von HT PEM und Reformer wurden identifiziert. Es wurde festgestellt, dass bei 180°C, eine HT PEMFC, die mit einer Celtec P® 2000 MEA versehen wurde, eine CO-Toleranz von bis zu 20% mit einem Verlust von etwa 150 mV (bei 200 mA/cm² der Last) ermöglicht, wohingegen die gleiche MEA nicht einmal 1% CO bei 130°C standhalten konnte (die Zelle verlor fast die gesamte Zellspannung, als sie mit 1% CO im H_2-Anode-Gasstrom bei 130°C belastet worden ist).

Eine HT PEMFC-Stack-Toleranz zu anderen synthetischen Reformaten (bestehend aus verschiedenen Wasserstoffkonzentrationen) wurde ebenfalls untersucht.

Die HT PEM-Brennstoffzellen-Degradation, die durch den Verlust von Phosphorsäure (Elektrolyt) und somit dem Verlust vom elektrochemisch aktiver Oberfläche (ECSA) verursacht wird, wurde untersucht.

ICP-OES und ICP-MS-Techniken wurden verwendet, um das Produkt-Wasser der Zelle zu analysieren, während Zahnermesstechnik's IM6 Analysator verwendet wurde, um die ECSA mittels CV (Cyclicvoltammetry) Scans zu studieren.

Basierend auf einem 1000-Stunden-Test wurde beobachtet, dass eine 10°C Temperaturerhöhung (von 160°C) zu einem Elektrolytenverlust von > 2,0-fachen seines Wertes bei 160°C führt. Der Elektrolyt Verlust lag bei etwa 0,45 µg /m² /s bei 170°C pro Zelle, während der wert bei 0,20 µg /m² /s bei 160°C pro Zelle lag.

Die Langzeit Performance (2000 Betriebsstunden der Zelle) und die beschleunigte Zell Degradation (beim Betrieb mit Null Laststrom oder OCV) wurden untersucht.

Die Widerstandsfähigkeit der Zellbestandteile wurde mit Hilfe von REM (Rasterelektronenmikroskop)-Bildern untersucht.

Nach dem OCV-Test (120 stunden) wurde beobachtet, dass der Verlust in der Zelle ECSA 20% vom Anfangswert war, der Spannungsverlust betrug 16,6% vom Anfangswert.

Aus dem 2000-Stunden-Test geht hervor, dass die Zelle 42,2 mV (0,6610 V bis 0,6188 V)(oder 6,4%) bei einer durchschnittlichen Belastung von 200 mA/cm², bei einer durchschnittlichen Zelltemperatur von 160°C verloren hat.

Das entspricht einem Performance-Verlust in Höhe von 10 µV/h auf konstanter Last (200 mA/cm²) und einem Durchschnitt von 20 µV/h, wenn in der Testphase konstante Last, Lastwechsel und Temperaturwechsel, Shutdown und Startup-Phase enthalten sind.

Table of Contents

Nomenclature

Chemical Formulae

Formulae	Unit	Details
H_3PO_4	moles	Phosphoric Acid
$H_2PO_4^-$	moles	Phosphoric acid anion
PO_4	moles/gram	Phosphate ion
H_2PO_4	moles	Dihydrogen phosphate
F = 96485	[C/mol]	Faraday's Constant
C_p	[kJ/kg*K]	Specific Isobaric heat capacity
$\Delta G°_{rxn}$	[J/mol]	Gibbs free energy of reaction
$\Delta H°_{rxn}$	[J/mol]	Enthalpy of Reaction
$\Delta S°_{rxn}$	[J/mol.K]	Entropy of Reaction
E_{th}	[Volts]	Maximum theoretical cell Voltage
$E°$	[Volts]	Ref. Cell Voltage (Ideal FC Voltage) at 25°C, 1atm
R = 8.314	[J/mol.K]	Gas Constant
η_{max}	[dimensionless]	Maximum thermodynamic efficiency of a fuel cell
I	[A]	Current
j or i	[mA/cm^2]	Current density

Abbreviations

Abbreviation	Details
BPP	Bipolar plate (in most cases this could mean bipolar half plate also)
CC	Current Collector plates
GDL	Gas Diffusion Layer
GDE	Gas Diffusion Electrode
MPL	Microporous Layer
CL	Catalyst Layer
RTP	Room Temperature and Pressure
PA	Phosphoric acid
LT PEMFC	Low temperature polymer electrolyte membrane fuel cell
HT PEMFC	High temperature polymer electrolyte membrane fuel cell
PBI	Polybenzimidazole
I.V.	Inherent Viscosity

Abbreviation	Details
ORR	Oxygen reduction reaction
MEA	Membrane Electrode Assembly
OCV	Open circuit voltage
ECSA	Electrochemically active surface area
MFC	Mass Flow Controller
$i_{crossover}$	Equivalent fuel crossover current density
i_L	Limiting current density
i_{0c}	Exchange current density (of cathode reaction)
i_{0a}	Exchange current density (of anode reaction)
C_{surf}	Surface concentration of species
C_{Bulk}	Bulk concentration of species
d_m	Membrane thickness
α	Electron transfer coefficient
η_{act}	Activation overvoltage
η_{iR}	Ohmic overvoltage
η_{conc}	Concentration overvoltage
$\eta_{crossover}$	Overvoltage due to fuel crossover through membrane
ΔP_{H2}	Differential partial pressure of H_2
ΔP_{O2}	Differential partial pressure of O_2
k_{H2}	Permeability coefficient for hydrogen
k_{O2}	Permeability coefficient for oxygen
TPB	Triple phase boundary
BOL	Beginning of Life
EOL	End of Life
CHP	Combined Heat and Power
q_{pt}	Charge density due to H-adsorption
H_{area}	H-adsorption curve area
HHV	Higher heating value
LHV	Lower heating value
AECD	Apparent exchange current density
W	Warburg impedance
CPE	Constant phase element
R_{ct}	Charge transfer resistance
C_{dl}	Double layer capacitance
$-Z_{imag(max)}$	Negative Maximum imaginary impedance from nyquist plot of EIS spectra

1 Motivation

According to the World Bank's estimates, the world's population in 2011 is around 7 billion. Higher population implies higher energy consumption and higher energy consumption implies higher air pollution, if appropriate measures are not taken. Fuel cells offer tremendous advantages such as energy conversion (from chemical energy of fuels to electrical and heat energy) with little or ultra low emissions. PEMFCs, PAFCs have been researched for the past four decades for use in terrestrial as well as extra terrestrial applications. DMFCs have been researched for portable applications. However, problems associated with water management, high cost of the membranes, durability of MEA components have limited the use of low temperature (LT) PEMFCs. However, there are a few companies such as Smart Fuel Cells in Germany developing commercially available DMFC based systems. As the electrolyte used in PAFCs is a liquid, most of its applications were based on stationary applications ranging up to some hundreds of kWs. These PAFCs were not suitable for small or medium size applications. DMFCs have some inherent problems such as fuel crossover and low conversion efficiencies. Therefore only little progress was made by these types of fuel cells and their associated systems in a commercial sense. MCFCs and SOFCs have become commercially viable, in the kW range for residential applications and as well as in large CHP (combined heat and power) units of hundreds of kWs. Whereas high temperature PEMFCs (HT PEMFCs), operating in the temperature range of 120°C – 200°C are rather new and are explored due to the advantage of their operating window. For instance fuel cells operating at > 100°C reduce issues related to water management substantially. Circulating excess heat energy from such fuel cells into other system processes where heat is needed would be much more practical (due to higher ΔT) compared to the standard LT PEMFCs where the produced heat has less than 90°C (lower ΔT). Higher tolerance to fuel impurities such as CO, by these HT PEMFCs has made them very practical for many applications. This is a very young technology. Although PBI/H_3PO_4 based membranes have been explored for use in PEMFCs from the early 1990s, only recently PEMEAS (currently BASF) has marketed them as commercially available MEAs. Besides, some companies such as Sartorius (currently Elcomax), Germany, Danish Power system, Denmark are offering MEAs on a commercial basis. Although some issues remain, such as development of durable and low cost catalyst and catalyst support materials, acid management, the rapid development of membranes and MEAs has been motivated by a huge demand from many a market. Recently, DLR in Germany has tested its pilot airplane (Antares) fully operated with a HT PEMFC stack (with on board H_2 bottle). ClearEdge Power in Portland, USA has been developing systems based on HT PEMFC technology to be deployed in the US as well as in South Korean households. Many more companies are increasingly interested in this technology due to the many fold advantages it has to offer.

This work is aimed at elucidating this HT PEMFC technology, in terms of giving an in-depth view of what it means to operate a HT PEMFC. Single cell characterization at different operating temperatures, analysis of impedance contributions from cell components, HT PEMFC's tolerance to fuel impurities, long term performance of HT PEMFCs, degradation phenomena observed in HT PEMFCs, acid (electrolyte) management, component-related aspects, kinetic parameters governed by higher temperature operation, OCV operation, have been explored and discussed in the current work. Although there exists a brief variety of high temperature specific MEAs from 5 groups (such as from DTU of Denmark, CWRU of USA, Advanced Technologies of Greece, Elcomax of Germany, Fuma Tech of Germany), the commercially available Celtec P® MEAs have been studied here, due to the reproducibility of cell performance and due to the possibility of procurement by any end user.

Theoretical values of maximum thermodynamically possible EMF of a HT PEMFC operating at temperatures such as 120°C, 130°C or 200°C and maximum theoretical thermodynamic efficiencies were derived from first principles (after deriving entropy and enthalpy values at various temperatures). As the existing literature is mostly relavent to LT PEMFCs and SOFCs, these derivations were necessary to accurately document efficiency and OCV (open

6

circuit voltage) maxima (limits) pertaining to HT PEMFCs. Also, lowly doped and highly doped HT-MEAs were compared and discussed in the current work. The overall aim of this work is to make explicit the pros and cons of this HT PEMFC technology with its long term use in mind and with a perspective that this technology might be one of the most preferred options of the current century, possibly giving rise to deployment of many HT PEMFC systems around the world.

2 Theoretical Background

Fuel cells can be categorized as depicted in Table 2.1, depending on their cell reactions. PEMFCs have protons (as their mobile ions), the AFCs have OH⁻ (hydroxyl ions) (as their mobile ions), MCFCs have carbonate ions (as their mobile ions), SOFCs have O^{2-} ions (as their mobile ions). Out of these, PEMFCs and PAFCs have drawn a lot of attention, due to their high volumetric and gravimetric power densities and being simple to operate. SOFCs are good competitors to these proton exchange types. But the lower temperature option (130°C – 200°C operation) of PAFCs compared to (650°C – 1000°C operation) of SOFCs, implying shorter start up times, has been desired in many applications. Therefore, PEMFCs and PAFCs have been preferred by many developers, due to their simple operation, simple construction and lower operating temperatures. However, there are some technical barriers to make these PEMFCs commercially viable as enumerated in the current work. After about 4 decades of research and development, the standard LT PEMFC (low temperature based PEMFC) technology is yet to make a commercial case. The durability, cost of membranes and catalysts, acceptable proton conductivity (> 0.1 S/cm), still need a further breakthrough. Against such a background, HT PEMFCs appear to be more promising, at least for stationary applications, just by virtue by their region of operation (130°C-200°C), making the total system simpler, lighter and practical (tolerating high amounts of CO which is normally present in reformates).

Table 2.1: Types of fuel cells and their reactions [39]

Types of Fuel Cells	Anode side Reaction	Cathode side Reaction
Proton Exchange Membrane (PEMFC & PAFC)	$H_2 \rightarrow 2H^+ + 2e^-$	$0.5O_2 + 2H^+ + 2e^- \rightarrow H_2O$
Alkaline (AFC)	$H_2 + 2(OH)^- \rightarrow 2H_2O + 2e^-$	$0.5O_2 + H_2O + 2e^- \rightarrow 2(OH)^-$
Molten Carbonate (MCFC)	$H_2 + CO_3^{2-} \rightarrow H_2O + CO_2 + 2e^-$ $CO + CO_3^{2-} \rightarrow 2CO_2 + 2e^-$	$0.5O_2 + CO_2 + 2e^- \rightarrow CO_3^{2-}$
Solid Oxide (SOFC)	$H_2 + O^{2-} \rightarrow H_2O + 2e^-$ $CO + O^{2-} \rightarrow CO_2 + 2e^-$ $CH_4 + 4O^{2-} \rightarrow 2H_2O + CO_2 + 8e^-$	$0.5O_2 + 2e^- \rightarrow O^{2-}$
e^- - electron	CO_3^{2-} - carbonate ion	H_2O - water
H^+ - hydrogen ion	CO_2 - carbon dioxide	O_2 - oxygen
OH^-- hydroxyl ion	CO - carbon monoxide	H_2 - hydrogen

Fuel cell technology has been the most preferred option from amongst the existing power conversion devices due to their high potential in terms of specific energy. The Ragone plot [1] of power density versus energy density of most of the well known power conversion technologies is depicted in Fig 2.1. When fuel storage is considered in the case of fuel cells, specific energies in the range of 325 – 390 Wh/kg has been demonstrated [2], which is still higher than what state-of-the-art batteries can deliver. For instance, a 25 W Ultracell® system operated with reformed methanol and a HTPEMFC stack weighs 1.14 kg without the fuel cartridge. When a 1.4 kg methanol cartridge (equivalent to 1000 Wh capacity) is included, the specific energy amounts to 393.7 Wh/kg [67].

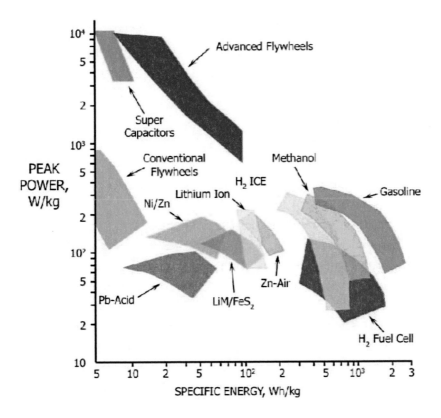

Figure 2.1: Ragone plot of power density versus energy density

As can be seen from the Ragone plot, fuel cells, once commercially viable hold a great promise when compared to the existing energy conversion technology options. Cost reduction (membranes and catalysts) and durability are still a major concern. Among the existing many types of fuel cell technologies, a small comparison is made between a standard LT PEMFC, HT PEMFC and SOFC in Table 2.2, taking a few criteria into account.

Table 2.2: Salient features of PEMFC, HT PEMFC and SOFC [3]

Criteria	PEMFC	HT-PEMFC	SOFC
Operating Temp.	25°C - 85°C	130°C - 200°C	600°C - 1000°C
Membrane Cost	$ 491-700/m² (Nafion based) [4,5]	10% of Nafion based ones ($1500/kW) [6]	10% of Nafion based
MEA Cost	$ 101/ kW [7,8]	$ 130/ kW	$92 /kW [9]
Reformer Size	Large (>CO in ppm)	Small (>CO in %)	Small (>CO in %)
Startup time	minutes (time to hydrate – humidifiers to work)	minutes (to heat up to 150°C)	hours (heat up to 700°C)
Water management	can be difficult	easy	easy
Gas feeds	need to be humidified	can be dry	can be dry
CO tolerance	20 - 50 ppm	1%(160°C) - 15%(180°C)	30% - 40% [10]
Operating Life	26,000 hours	18,000 hours (BASF)	30,000 hours (NEDO)

With regard to HT PEMFC development, apart from catalyst and catalyst supports, microporous layers and gas diffusion layers, membrane development has been the focus of many groups as highlighted here. Several classes of membrane materials for operation at elevated temperature have been investigated till date. Kerres [11], Meier-Haack [12] and Smitha et al [13] have reviewed the state-of-the-art in membranes for PEMFCs. As alternatives to Dupont's Nafion®, Flemion® membranes were developed by Asahi Glass Company and Aciplex® membranes were developed by Asahi Kasei of Japan, which could operate at about 120°C. Subsequently, several attempts were made to realize Nafion® composites which would operate in the medium temperature range (120°C – 150°C). The first approach was to improve the water retaining properties of state-of-the-art perfluorinated materials by incorporation of hydrophilic inorganic materials, like silica, zirconium phosphate or ZrO_2 and other phosphates and oxides or also polyacids as inorganic fillers. Mauritz et al [14] have developed Nafion-SiO_2 and Nafion-ZrO_2 composites. Alberti and Casciola [15] have reviewed the state-of-the-art of these composite membranes for use in medium temperature PEM fuel cells. These composite membranes are developed for operation in fuel cells at temperatures above 100°C or for improved DMFC with reduced methanol crossover. At the same time, advances made in the improvement of the perflourinated polymer backbone led to its improved temperature and long term stability, although requiring liquid water as a proton transfer medium.

Consequently, several groups [16,17] have started to develop non-fluorinated membranes which would operate at temperatures beyond 130°C. Kreuer [18] investigated polyether-etherketones (PEEK) and blended them with polymers containing immobilized heterocycles such as imidazole, benzimidazole or pyrazole which would function as proton solvents. This effort by Kreuer had heralded the strive for anhydrous (conductivity not critically dependant on water content) proton conductors.

Phosphoric acid had first appeared to be a promising candidate to be used as proton solvent at temperatures of 130°C to 200°C. Attempts were made to dope H_3PO_4 in basic polymers such as PBI (Polybenzimidazole).

PBI is an amorphous [19,20], basic polymer with a glass transition temperature close to 430°C [21]. Since the 1980s, many polymer blends have been explored which could be doped with either phosphoric acid or sulfuric acid to facilitate proton conductivity. Donoso et al [22] explored PEO/H_3PO_4 (Poly ethyleneoxide-phosphoric acid) complexes, Daniel et al [23] have explored PEI/H_3PO_4 (Polyethyleneimine-phosphoric acid) complexes. Schuster and Meyer [24] have reviewed these anhydrous polymer/acid complexes in detail. For the first time, PBI/H_3PO_4 blend was explored as a possible anhydrous polymer membrane material for use in PEMFCs and DMFCs by Wainright and Savinell [25] in 1995. Following Wainright and Savinell's efforts, extensive research was performed by many which have resulted in various types of PBI/H_3PO_4 systems, some of which will be described in the following paragraphs. Furthermore, as an alternative to PBI/H_3PO_4 systems, fully polymeric systems which exhibit protonic conductivity as an intrinsic property by forming mobile protonic defects such as excess protons or proton vacancies have been envisaged. For example, Polystyrene and polysiloxane architectures were considered as a possible route to achieving fully polymeric materials by Herz et al [26]. However, Schuster and Meyer have argued that the proton conductivities of such systems need improvement to be considered viable. Referring to the PBI/H_3PO_4 systems, followed by their initial success, many groups (see Fig 3.18) have started their research work to realize anhydrous proton conducting membranes for use in PEMFCs at elevated temperatures.

These PBI/H_3PO_4 based systems have exhibited significantly differing proton conductivities based on casting/preparation method employed. Fig.3.18 in Chapter 3 depicts the popular acid/base systems that have been developed recently and appear to be very promising. To elucidate, Savinell's group had blended TFA (trifluoroacetic acid) with prestine PBI and cast them into membranes, which were later immersed into aqueous phosphoric acid. Resorting to this method, they achieved conductivities close to 0.009 S/cm at 200°C [27]. Conductivity of these membranes was further improved by slightly increasing the water activity or water partial pressure. About 0.02 S/cm to 0.04 S/cm at 190°C was achieved with 0.5 to 1.5 bar of water partial pressure, when every mole of PBI repeat unit was doped with 5 moles of H_3PO_4 [25]. However, those membranes cast from TFA/PBI blends exhibited very poor mechanical properties and were difficult even to fabricate into MEAs [28]. The mechanical integrity of the polymer conducting membrane in a MEA is critically important because it serves as a mechanical support for electrodes and also as a barrier for oxidant and fuel (preventing possible mixing). In addition, it must be possible to fabricate MEAs (membranes with electrodes and microporous layers in some cases). The same group had also cast high temperature stable membranes by dissolving commercially available prestine PBI with DMAc (Dimethyl Acetamide) [29]. DMAc is an organic solvent which can dissolve PBI completely. It was observed by them that the polar groups in PBI had interacted with DMAc very strongly and even after very laborious treatment to remove trace amounts of DMAc from the membrane, some trace amounts of DMAc found their way into MEAs. And in those MEAs, those trace amounts of DMAc were found poisonous to the platinum based catalyst [30,31]. However, these membranes had shown higher proton conductivities compared to those developed by TFA blending process. For instance, PBI cast from DMAc had conductivities of 0.01 to 0.04 S/cm in the temperature window of 130°C to 190°C [32]. Furthermore, these DMAc cast membranes had higher mechanical strength compared to those cast from TFA process, although the doping level (H_3PO_4/PBI) of these two types tested were the same.

Li and Bjerrum [33,34] have doped the commercially available PBI with about 13 – 16 moles of PA (phosphoric acid) per mole of PBI repeat unit and had achieved a proton conductivity of 0.13 S/cm at 160°C. Although they had reported their highest doping level as 16 moles of PA per mole of PBI repeat unit, their membranes had exhibited very poor mechanical properties at doping levels of > 10 moles of PA per one mole of PBI repeat unit. Following these attempts, Xiao and Benecewicz [35] have synthesized PBI of high inherent viscosities in PPA (Polyphosphoric acid). PPA was used as a polycondensation agent and as a PBI solvent. They had analysed various PBI copolymers with varying polymer backbone rigidities using this process. For instance, various pyridine based PBI homopolymers with additional main chain pyridine groups were synthesized from TAB (tetraaminobiphenyl) and pyridine DBAs (dicarboxylic acids) using PPA. The polymerization solution was hydrolysed to produce highly doped membranes. Following their work, series of highly doped membranes (H_3PO_4/PBI) were prepared and commercialized by PEMEAS (currently BASF fuel cells). These membanes are marketed as Celtec P® 1000 or Celtec P® 2000 or Celtec P® 2100 type of MEAs from BASF fuel cells. The doping level of these membranes can be anywhere between 20 – 32 moles of PA per mole of PBI repeat unit, as elucidated by Xiao et al, [81]. These membranes with very high doping have exhibited unprecedented levels of proton conductivity. For instance, at 160°C, they had proton conductivity of > 0.22 S/cm and at 200°C, a proton conductivity of > 0.25 S/cm. This gave rise to the commercial success of these highly doped gel based membranes.

2.1 Fuel Cell Thermodynamics

Thermodynamics of fuel cells serve as a necessary foundation to understand and appreciate various fuel cell reactions. For instance, the maximum cell voltage that can be expected from a fuel cell fed with a given type of oxidant and fuel at a given temperature and pressure has a theoretical maximum, governed by thermodynamics. It is important to know these theoretical limits of cell voltage at different operating temperatures of the fuel cell type in question, to understand the cell's conversion efficiencies at various temperatures. Although there exist standard thermodynamic tables, which give information on enthalpies, entropies and Gibbs free energy values for various species, detailed and specific data on theoretical reversible potential and theoretical maximum cell efficiencies of cell operating in the temperature window of 100°C-200°C is rare to find in open literature. In the current work, the HT PEMFC's theoretical thermodynamic voltage (E_{th}) at various temperatures (100°C – 230°C) and the maximum attainable theoretical efficiencies (η_{max}) at these temperatures are derived from first principles. Fig 2.2 depicts various performance losses occurring in a fuel cell from the mentioned theoretical thermodynamic limit, which can be found in any standard text book.

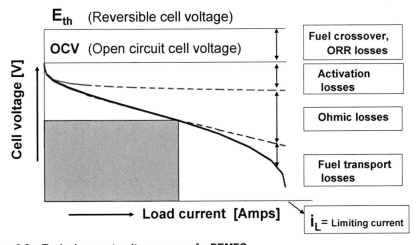

Figure 2.2: Typical current-voltage curve of a PEMFC

2.1.1 Enthalpy, Entropy and Gibbs free energy

For PEM Fuel Cells operating at temperatures from 25°C to 200°C (LT PEM and HT PEM), it is necessary to know the enthalpy, entropy and Gibbs free energy of cell reaction at various cell operating temperatures. A good understanding of these parameters helps understand the maximum theoretical cell voltage (E°) and the maximum theoretical cell efficiencies (η_{max}) of PEMFCs operating in the mentioned temperature regions. This section gives an overview of these essential parameters. From the enthalpy, entropy and Gibbs free energy values (for different species such as hydrogen, oxygen and product water) at reference temperature (25°C, 1 atm), it is possible to calculate enthalpy, entropy and Gibbs free energy values at temperatures higher than 25°C [36]. In the current work, these values at T_2 (25°C to 230°C) were calculated. From these calculated values at T_2, for a fuel cell reaction in a PEMFC where one mole of hydrogen reacts with half a mole of oxygen to produce one mole of water, the values of net Enthalpy, net Entropy and net Gibbs free energy for the said fuel cell reaction were calculated. From these values, the maximum theoretical thermodynamic potential and maximum cell efficiency of a fuel cell operating with H_2/O_2, producing water, were calculated at T_2 (25°C to 230°C). The calculated values can be found in tables shown in Appendix I (Table 13 at the end of the current work). The relations shown in (Eq.2 through to Eq.6), are the governing principles for calculating various quantities of Enthalpy, Entropy and Gibbs free energy, maximum theoretical cell voltage and maximum theoretical cell efficiency at any temperature above the standard reference temperature of 25°C, as enumerated further.

The reversible potential of a fuel cell at temperature T_2 is (effects of reactant and product partial pressures not taken into account here) given by the relation shown (Eq.1).

[Reversible Cell Voltage at $T_2°$ K]	$E = \dfrac{-\Delta G°_{rxn}}{nF}$	Eq. 1

Where $\Delta G°_{rxn}$ is the Gibbs free energy of reaction, "n" is the number of electrons participating in the reaction and "F" is Faraday's constant. This Gibbs free energy at T_2 is given by the relation shown in (Eq.2), where $\Delta H°_{rxn}$ is the net Enthalpy of cell reaction at T_2 and $\Delta S°_{rxn}$ is the net Entropy at the same temperature.

[Gibbs free energy of reaction at $T_2°$ K]	$\Delta G°_{rxn} = \Delta H°_{rxn} - T_2 \cdot \Delta S°_{rxn}$	Eq. 2

The values of $\Delta H°_{rxn}$ and $\Delta S°_{rxn}$ at various temperatures could be calculated using the relations shown in (Eq.3a, Eq.3b, Eq.4a and Eq.4b). Subsequently, the value of $\Delta G°_{rxn}$ at any temperature above the reference temperature (25°C or 298.15°K) could be computed by taking the corresponding $\Delta H°_{rxn}$ and $\Delta S°_{rxn}$ values at that temperature and using them in the relation shown in (Eq.2).

Net enthalpy of reaction ($\Delta H°_{rxn}$) at any temperature can be calculated by using the relation shown in (Eq.3a) where water is the product, hydrogen and oxygen are reactants. Further, the enthalpy of different species considered here (namely hydrogen, oxygen and water) at any temperature T_2 can be calculated using the relation shown in (Eq.3b). Where C_p is the heat capacity of the respective species considered and $\Delta H° (T_1)$ is the standard enthalpy of that species at 25°C or 298.15°K.

Enthalpy of reaction at $T_2°$ K	$\Delta H°_{rxn} = \Delta H° (H_2O) - \Delta H° (H_2) - 0.5 * \Delta H° (O_2)$	Eq.3a
Enthalpy of species at $T_2°$ K	$\Delta H°_{T2} = \Delta H°\left(T_1\right) + \int\limits_{298.15°K}^{T2} C_p \, dT$	Eq.3b

Similarly, the net entropy of a PEMFC reaction and the entropy of species at any temperature T_2 can be calculated using the relation shown in (Eq.4a and Eq.4b) where $\Delta S° (T_1)$ is the standard entropy of that species at 25°C or 298.15°K and C_p is the heat capacity of that species.

Entropy of reaction at $T_2°$ K	$\Delta S°_{rxn} = \Delta S° (H_2O) - \Delta S° (H_2) - 0.5 * \Delta S°(O_2)$	Eq. 4a
Entropy of spcies at $T_2°$ K	$\Delta S°_{T2} = \Delta S°\left(T_1\right) + \int\limits_{298.15°K}^{T2} \frac{1}{T} C_p \, dT$	Eq. 4b

As heat capacity is a function of temperature, C_p can be calculated using the empirical relations shown in (Eq.5 and Eq.6). The values of A,B,C and D could be obtained from standard thermodynamic tables [36,37] and are shown in Appendix I (Table 4).

For Liquids	$\dfrac{C_p}{R} = A + BT + CT^2$	Eq.5
For ideal gases	$\dfrac{C_p}{R} = A + BT + CT^2 + DT^{-2}$	Eq.6

2.1.2 Ideal Fuel Cell Voltages

Ideal cell voltage or the maximum theoretical voltage of a PEMFC at any operating temperature can be calculated as explained in the section 2.2.1 using the relation shown in (Eq.7).

[Reversible Cell Voltage at $T_2°K$]	$$E_{th} = \frac{-\Delta G°_{rxn}}{nF}$$	Eq. 7

In the current work, the $\Delta G°_{rxn}$ values at any cell temperature T_2 were calculated and are shown in Appendix I (Table 13). From these values, the ideal cell voltage was then calculated by using the standard values for "n" and "F". It can be concluded that, for PEMFCs operating in the 25°C – 90°C temperature region, where the product water is assumed to be in the liquid form, the E_{th} (or the maximum theoretical cell voltage) based on higher heating value (HHV) can be expressed by the relation shown in Eq.8.

[E_{th} (298.15°K – 363.15°K)]	= (1.47600 -0.8299E-3) * T (T is the cell temperature in °K)	Eq.8

Figure 2.3 depicts the E_{th} at various cell operating temperatures from 25°C to 90°C (or 298.15°K to 363.15°K).

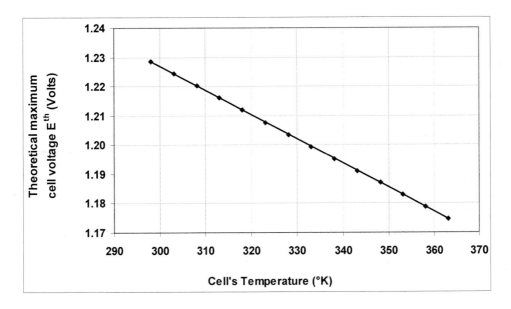

Figure 2.3 : E_{th} at operating temperatures from 25°C to 90°C (or 298.15°K to 363.15°K)

Similarly, for PEMFCs operating in the 100°C – 230°C (or 373.15°K to 503.15°K) tempera-ture region, where the product water is assumed to be in the vapour form, the E_{th} (or the maximum theoretical cell voltage) is a bit different, as the reaction enthalpy, entropy are dif-ferent under these conditions of cell operation. From the calculated values of net reaction en-thalpies, net entropies at T_2 (see Appendix I, Table 13), the maximum theoretical cell voltage for PEMFCs operating at T_2 (100°C-230°C) based on HHV can be expressed by the relation shown in Eq.9.

$[E_{th} (373.15°K – 503.15°K)]$	= (1.259266 -0.2466E-3) * T	
		Eq.9
	(T is the cell temperature in °K)	

Figure 2.4 depicts the E_{th} at cell operating temperatures from 100°C to 230°C (or 373.15°K to 503.15°K).

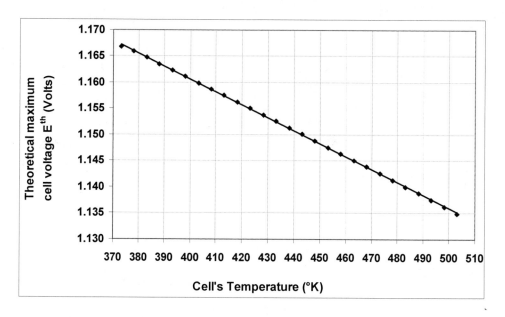

Figure 2.4: E_{th} at operating temperatures from 100°C to 230°C (or 373.15°K to 503.15°K)

Figures 2.3 & 2.4 show that the maximum theoretical fuel cell voltages are lower at higher cell operating temperatures. From the relations shown in Eq.8 and Eq.9, it can be said that a voltage loss of about 0.845 mV to 0.815 mV per 1° rise in operating temperature (in the temperature range 25°C to 90°C) is experienced by a PEMFC. Similarly, about 0.172 mV to 0.254 mV per 1° rise in cell temperature (in the temperature range100°C to 230°C) is experienced by a PEMFC. To enumerate further, for a LT PEMFC operating at 50°C, the E_{th} is 1.207 V, where as for a HT PEMFC operating at 160°C, the same is 1.152 V (a difference of 55 mV). Also, lower E_{th} values also imply lower η_{max} (maximum theoretical fuel cell efficiency) under same operating conditions.

2.1.3 Ideal Fuel Cell Efficiencies

Ideal theoretical maximum thermodynamic efficiency of an electrochemical cell sets the limit for the attainable fuel cell efficiency and is directly related to the ideal cell voltage in that state (partial pressures of reactants and products not being considered here). In the previous section, it was shown that at higher fuel cell operating temperatures, the ideal fuel cell voltage falls. This implicitly means, the ideal theoretical fuel cell efficiency (η_{max}) also falls at higher cell operating temperatures (higher than the standard reference temperature, 25°C or 298.15°C). Typically η_{max} can be calculated using the relation shown in Eq. 10.

$[\eta_{max}$ at $T_2°K]$	$= \dfrac{\Delta G°_{rxn}}{\Delta H°_{rxn}}$ X 100	Eq. 10

Where $\Delta G°_{rxn}$ at T_2 is the net Gibbs free energy of fuel cell reaction at temperature T_2. Standard reference enthalpy of fuel cell reaction at T_1 (or 298.15°K) is represented by ($\Delta H°_{rxn}$ at T_1).

Again, from the calculated values (shown in Appendix I, Table 13) of net reaction enthalpy, and the net Gibbs free energy, the η_{max} values for a PEMFC at temperature T_2 (ranging from 25°C to 230°C) were calculated using Eq. 10. The same procedure was repeated at different temperatures (25°C – 230°C) and the resulting values are shown in Appendix I, Table 13. From these calculated cell efficiency values, it can be said that for PEMFCs operating in the 25°C – 90°C temperature region, where the product water is in the liquid form, the η_{max} (or the maximum theoretical cell efficiency) can be expressed by the relation shown in Eq. 11 (see Fig 2.5). To enumerate further, Figure 2.5 below depicts the η_{max} at cell operating temperatures from 25°C to 90°C (or 298.15°K to 363.15°K).

$[\eta_{max}$ (298.15°K – 363.15°K)]	$= (0.99648 -0.56E-3) \cdot T$ (T is the cell temperature in °K)	Eq.11

19

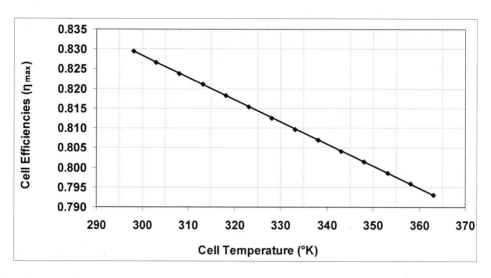

Figure 2.5 : η_{max} at operating temperatures from 25°C to 90°C (or 298.15°K to 363.15°K)

Following the same method, from the values shown in Appendix I (Table 13), the maximum theoretical cell efficiencies were calculated for a PEMFC operating at T_2. And it can be concluded that for PEMFCs operating in the 100°C – 230°C temperature region, where the product water is in the vapour form, the η_{max} (or the maximum theoretical cell efficiency) can be expressed by the relation shown in Eq.12 (see Fig 2.6).

$[\eta_{max} (373.15°K - 503.15°K)]$	= (0.85015 – 0.166E-3) * T (T is the cell temperature in °K)	Eq.12

Figure 2.6 depicts the η_{max} at cell operating temperatures from 100°C to 230°C (or 373.15°K to 503.15°K). Now, by looking at the η_{max} values at any cell temperature T_2, as shown in Eq.11 and Eq.12, it can be concluded that about 0.053% to 0.057% loss from its initial efficiency of 82.9 % (at standard reference temperature of 25°C and 1 atm) is experienced by a PEMFC for every 1° rise in cell temperature (in the cell temperature range 25°C to 90°C) and similarly, about 0.0161% to 0.0171% loss per every 1° rise in cell temperature (in the cell temperature range 100°C to 230°C) is experienced by the fuel cell.

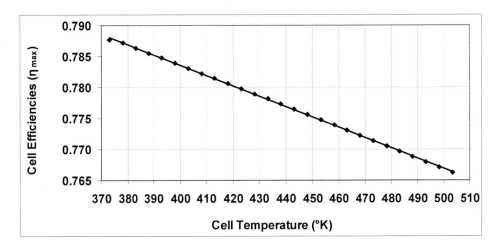

Figure 2.6: η max at cell temperatures of 100°C to 230°C (or 373.15°K to 503.15°K)

That implies that the rate of fall in theoretical cell efficiency is higher for LT PEMFCs when the cell temperature is raised from 25°C to 100°C, where as in HT PEMFCs (100°C to 230°C), the rate of fall (of theoretical cell efficiency) is lower. It can be seen from values shown in Appendix I (Table 13), that for a PEMFC operating at 25°C, the maximum thermo-dynamic efficiency is 82.9%, where as it is 81.5% at 50°C and it is 79.9% at 80°C. At 160°C, the efficiency is 77.8% and at 180°C, the same is 77.5%.

2.1.4 Nernst Equation

Nernst equation [38] for a fuel cell reaction could be used to express the reversible potential of a fuel cell as function of pressure or concentration of reactants.

For a fuel cell reaction of the type shown in the relation Eq.13, where A and B are reactants, C and D are products, with their respective coefficients (a,b,c,d), the change in Gibbs free energy could be expressed as shown in the relation Eq.14 [39], and can be found in standard text books.

$a*[A] + b*[B] = c*[C] + d*[D]$	Eq.13
$\Delta G = \Delta G^\circ + R*T*\ln\dfrac{[C^c]*[D^d]}{[A^a]*[B^b]}$	Eq.14

Also, the maximum electrical work that could be attained by a fuel cell operating at constant temperature and pressure is given by the relation shown in Eq.15, where ΔG_{rxn} is the change in Gibbs free energy of the fuel cell's electrochemical reaction, "n" denotes the number of electrons transferred in the reaction, "F" is Faraday's constant and E_{th} is the ideal cell potential.

$W_{el} = \Delta G_{rxn} = -nFE_{th}$	Eq.15

Also, if we take the reference of an electrochemical cell operating in the standard state (298.15°K and 1 atm), then the relationship between Gibbs free energy change and the cell's ideal potential will take the form shown in relation Eq.16, where E° stands for the ideal cell voltage or potential at standard reference temperature and pressure.

$W_{el} = \Delta G^\circ_{rxn} = -nFE^\circ$	Eq.16

Now, as the Eq.15 and Eq.16 represent the maximum electrical work attained by a fuel cell, they could be equated to each other. Therefore, by combining Eq.15 and Eq.16, we get the relation $(G_{rxn})/E = \Delta G^\circ_{rxn}/E^\circ$ and from this relation, the value of ΔG°_{rxn} can be expressed as $\Delta G^\circ_{rxn} = (\Delta G_{rxn})*E^\circ/E$. Now, substituting the ΔG° value in Eq.14 with ($\Delta G*E^\circ/E$) would yield the relationship for ideal cell voltage shown in Eq.17.

22

$$E = E° + \left[\frac{RT}{nF}\right] * \ln \frac{[A^a]*[B^b]}{[C^c]*[D^d]}$$

Eq.17

Furthermore, in an electrochemical cell, typically ΔG and ΔH are related to each other by the relation shown in Eq.18. The amount of heat that is produced in a fuel cell at any temperature T is given by (T·ΔS). Fuel cell reactions that have negative entropy change generate heat (meaning give out heat to the surroundings), whereas the fuel cell reactions with positive entropy change may extract heat from their surroundings if the irreversible generation of heat is smaller in quantity than the reversible absorption of heat.

[Gibbs free energy of reaction at T_2°K]	**ΔG° = ΔH° - T· ΔS**	Eq.18

Now, differentiating Eq.18 with respect to temperature gives rise to the relation shown in Eq.19:

At constant pressure, $\left[\dfrac{\partial E}{\partial T}\right] = \dfrac{\Delta S}{nF}$

Eq.19

Similarly, differentiating Eq.18 with respect to pressure gives rise to the relation shown in Eq.20:

At constant temperature, $\left[\dfrac{\partial E}{\partial P}\right] = -\dfrac{\Delta Volume}{nF}$

Eq.20

Summarizing the relations shown in Eq.13 to Eq.20, we obtain the relation for the thermodynamic open circuit voltage shown in Eq.21 here, which is a well known equation discussed in many fuel cell related text books, research papers and the like. The notations P_{H2}, P_{O2} and P_{H2O} denote partial pressures of fuel hydrogen, oxidant and the product water respectively. T is the cell temperature, R is gas constant, F is Faraday's constant. E° is the reversible potential of an electrochemical cell at standard temperature and pressure, which is typically 1.229 Volts when the product water is in the liquid form and it is 1.16 Volts when the product water is in the vapour form.

$E_{OCV} = E^\circ + \left[\dfrac{\Delta S}{n*F}\right]*[T - 298.15] + \left[\dfrac{R*T}{2*F}\right]*\ln\dfrac{[P_{H2}{}^{1}]*[P_{O2}{}^{0.5}]}{[P_{H2O}{}^{1}]}$	Eq.21 (a)
$E_{ocv} = 1.229 - [8.4692 - (5.017E-07)*(T-298.15)]*[T-298.15] + [4.3084E-05*T]*\ln\dfrac{[P_{H2}{}^{1}]*[P_{O2}{}^{0.5}]}{[P_{H2O}{}^{1}]}$	Eq.21 (b)
$E_{ocv} = 1.166 - [2.4185E-04 + (1.1656E-07)*(T-298.15)]*[T-298.15] + [4.3084E-05*T]*\ln\dfrac{[P_{H2}{}^{1}]*[P_{O2}{}^{0.5}]}{[P_{H2O}{}^{1}]}$	Eq.21 (c)

The relation shown in Eq.21 (a) represents the cell's open circuit voltage and is referred as the "Nernst Equation for PEM Fuel Cell reaction", where H_2 and O_2 are the reactants and water is the product. Based on the entropy values calculated and shown in Table 13 (Appendix I), Eq. 21(b) and Eq. 21(c) were derived. Eq. 21(b) is valid for cell operating temperatures of 25°C to 90°C and Eq.21 (c) is valid in the 100°C – 230°C temperature range. It may be noted here that typically the cell potential increases with an increase in the activity (or concentration) of the reactants and decreases with an increase in the activity of products. The second part in the Eq.21 accounts for the operating temperature of the cell and the third part accounts for the partial pressures of the reactant species. For instance, when the oxidant is air, as the partial pressure of oxygen is about 1/5th in air compared to a case of pure oxygen feed, the third part in Eq.21 will be lower, implying lower OCV compared to the case of pure oxygen feed. For instance, when air is used in stead of pure oxygen in the fuel cell reaction considered here, a loss of 15.2 mV at 180°C and a loss of 13.5 mV at 130°C can be expected as a result of reduced O_2 partial pressure. Similarly, when the fuel supplied is a reformate where the hydrogen partial pressure is only a fraction of the fed reformate's pressure, the cell voltage would be lower compared to the case of pure hydrogen feed. The Nernst equation for a few types of fuel cell reactions are listed in Table 2.3.

Table 2.3: Nernst equations for fuel cell reactions of various fuels [39]

Fuel cell reaction	Nernst equation for fuel cell voltage
$H_2 + 0.5 \, O_2 = H_2O$	$E = E° + (RT/2F)*[\ln(P^1_{H2}/ \, P^1_{H2O}) + \ln(P^{0.5}_{O2})]$
$H_2 + 0.5 \, O_2 + CO_{2c} = H_2O + CO_{2a}$	$E = E° + (RT/2F)*[\ln(P^1_{H2}/ \, (P^1_{H2O} \cdot P^1_{CO2a})) + \ln(P^{0.5}_{O2} \cdot P^1_{CO2c})]$
$CH_4 + 2O_2 = 2H_2O + CO_2$	$E = E° + (RT/8F)*[\ln(P^1_{CH4}/ \, (P^2_{H2O} \cdot P^1_{CO2})) + \ln(P^2_{O2})]$
$C + 0.5 \, O_2 = CO_2$	$E = E° + (RT/2F)*[\ln(P^1_{c}/ \, P^1_{CO2}) + \ln(P^{0.5}_{O2})]$
CO_{2a} = anode side CO_2 ; CO_{2c} = cathode side CO_2 ; P = Partial pressure of gases (atm); T in °K ;	
R = Gas constant; E = Nernst equation for fuel cell reaction.	

2.1.5 Performance limiting losses in PEM fuel cells

Eq.21(a-c) show the open circuit or the no load voltage of a typical PEMFC. However, to arrive at the actual performance (or the cell voltage) under load, generally we must take four types of losses into account, namely i) activation overvoltage, ii) ohmic overvoltage, iii) concentration overvoltage and iv) fuel crossover overvoltage. Fuel crossover loss usually is negligible (except for a DMFC). The actual potential or the voltage of a cell can be obtained by deducting these four losses from the ideal cell voltage (or the Nernst cell voltage discussed above).

2.1.6 Activation overvoltage

The cell voltage loss caused due to activation overvoltage (Eq.22) can be expressed using **Tafel equation** [40] which is a simplification of Butler-Volmer equation and is shown in Eq.23. The terms "α" and "i_0" denote the electron transfer co-efficient of the electrochemical reaction at the respective electrode (anode or cathode) and the exchange current density (representing rate of electrode reaction) respectively. "i" is the cell's load current density. R, T, F and n have their usual meaning.

$$\eta_{act} = \left[\dfrac{RT}{\alpha nF}\right] * \ln\dfrac{[i]}{[i_0]}$$		Eq.22
[Tafel equation]	$\eta_{act} = b * \ln\dfrac{[i]}{[i_0]}$ where b = Tafel slope	Eq. 23

From the first part of the Eq.23, "b" is called the Tafel slope and can be obtained from the slope of a plot drawn with η_{act} versus log (i) (the logarithm of fuel cell current density). This loss due to activation overvoltage, in essence takes into account the slow electrode kinetics, the processes involving absorption of reactant species, desorption of product species, transfer of electrons across the double layer and also the very nature and type of the electrode surface involved. The key kinetic parameters "α" and "i_0" for a HT PEMFC operating in the 130°C – 180°C temperature range were deduced from the experimental work carried out in the course of this work, which can be found in section 3.1.7 and 3.1.8.

2.1.7 Ohmic overvoltage

It is the cell voltage loss caused due to resistance of cell components and interfaces. The essential components of the fuel cell such as bipolar plates (BPPs), current collectors (CCs), gas diffusion electrodes (GDEs), the proton conducting membrane, all contribute to ohmic overvoltage, depending on their material composition and cell compression enumerated further in section 3.1.1.1. The PEMFC's ohmic overpotential is expressed as $(\eta_{iR} = i * R)$. The high frequency impedance contribution from BPPs, CCs and GDEs is usually referred to as electronic resistance and the same from the fuel cell membrane and the ionomer present in the catalyst layer, GDL, microporous layer is referred as ionic resistance. The performance loss due to ohmic overvoltage in a HT PEMFC is discussed based on the experiments conducted in the course of this work and are documented in section 3.1.1.

2.1.8 Diffusion Overvoltage

Thirdly, the cell voltage loss caused due to concentration (or diffusion or fuel transport) over-voltage accounts for the rate of mass transport into the reaction region (or the triple phase boundary or TPB regions). This is usually described by Fick's first law [41,42,43] of diffusion as shown in Eq.24 [39].

$$\left[i_{conc} \right] = -\frac{D*n*F*(C_{Bulk} - C_{Surf})}{d} \qquad \text{Eq.24}$$

Where

D = Diffusion coefficient of the reacting species

n = number of electrons

F = Faraday's Constant

C_{Bulk} = Bulk concentration

C_{Surf} = Surface concentration

d = Diffusion layer thickness

i_{conc} = represents rate at which species are transported to the reaction region

Furthermore, the maximum rate of reactant supply to an electrode, while drawing load current, under given conditions of fuel feed stoichiometry, temperature and pressure of operation, also called as the *limiting current* (i_L) can be expressed as shown in Eq.25, as it occurs when C_{Surf} (surface concentration) virtually falls to zero.

$$\text{Limiting current, } \left[i_L \right] = -\frac{n*F*D*C_{Bulk}}{d} \qquad \text{Eq.25}$$

By combining Eq.24 and Eq.25, the relation shown in Eq.26 is obtained.

$$\frac{C_{Surf}}{C_{Bulk}} = 1 - \left[\frac{i}{i_L}\right]$$

Eq.26

Under no load (or OCV) conditions, typically the Nernst equation for the reacting species could be expressed as shown in Eq.27.

$$E_{i=0} = E° + \left[\frac{RT}{nF}\right] * \ln[C_{Bulk}]$$

Eq.27

When some load current is being drawn from the electrochemical cell, typically as the surface concentration of the reacting species becomes much less than the bulk concentration, the Nernst equation of the electrochemical reaction takes the form shown in Eq.28.

$$E_{i \supset 0} = E° + \left[\frac{RT}{nF}\right] * \ln[C_{Surf}]$$

Eq.28

The concentration overvoltage, which is the potential difference caused by a change in concentration of species at the electrodes can be expressed as shown in Eq.29.

$$\eta_{conc} = \Delta E = \left[\frac{RT}{nF}\right] * \ln\left[\frac{C_{Surf}}{C_{Bulk}}\right]$$

Eq.29

Combining Eq.26 and Eq.29 yields us the Eq.30, which represents cell's loss caused due to diffusion overvoltage.

$$\eta_{fueltransport} = \Delta E = \left[\frac{RT}{nF}\right] * \ln\left[1 - \frac{i}{i_L}\right] \qquad \text{Eq.30}$$

However, Eq.30 is a simplified expression for fuel transport overvoltage which does not include fuel crossover loss, which can be neglected in the case of cells operated with relatively higher gas feed stoichiometries ($\lambda=1.35$ for H_2 and $\lambda=2.5$ for air in this work) and are operated in the lower (20 – 400 mA/cm²) load current range. Also, CPE (constant phase element) or Warburg impedance type behaviour (see section 3.1.1) is common in HT PEMFCs, which contributes to gas diffusion limitations even at lower currents. This is due to their inherent drawback of gas diffusion limitations into reaction sites (TPB) due to the presence of electrolyte barrier. However, at the mentioned high gas feed stoichiometries, these diffusion loses are not significant. Therefore, the cell performance loss due to concentration overvoltage in a HT PEMFC was calculated using Eq.30, based on the experimental work carried out in the course of this work and is enumerated further, in section 3.1.3.

2.1.9 Fuel crossover loss

Permeation of reactant gases through fuel cell membranes causes fuel cell crossover efficiency loss [44,45] which has two components: a) Loss due to direct electrochemical reaction of permeated hydrogen at the cathode and b) Loss due to direct reaction of permeated oxygen at the anode. In DMFCs, methanol crossover resulting in mixed potentials is a major challenge. This loss can be expressed as an equivalent current that would be observed externally, had the reactant gases lost due to crossover, been used in the standard fuel cell reaction and is shown in Eq.31, where $i_{crossover}$ is the fuel crossover current density. $\eta_{crossover}$ is the fuel crossover overpotential (Eq.32).

$$i_{crossover}\ (A/cm^2) = \left[\frac{2 * F * k_{H2}(T,P) * \Delta P_{H2}}{d_m}\right] + \left[\frac{4 * F * k_{O2}(T,P) * \Delta P_{O2}}{d_m}\right] \qquad \text{Eq.31}$$

$$\eta_{crossover} = \left[\frac{RT}{\alpha nF}\right] * \left[\ln\frac{[i + i_{crossover}]}{[i_0]} - \ln\frac{[i]}{[i_0]}\right] \qquad \text{Eq.32}$$

k_{H2} (T,P), k_{O2} (T,P) = Permeability coefficient for H_2 and O_2 respectively, through the fuel cell membrane material (mole.cm/cm².s.kPa)

ΔP_{H2} , ΔP_{O2} = Differential partial pressure of H_2 and O_2 respectively, across the membrane and is the driving force for gas permeation (kPa)

d_m = membrane thickness (cm)

i = Load current density (mA/cm²)

i_0 = Exchange current density (mA/cm²)

$i_{crossover}$ = Equivalent fuel crossover current (mA/cm²)

Furthermore, gas permeation in high temperature stable PBI/H_3PO_4 based membranes are discussed with more details in section 2.2.1.

Plug Power Inc., USA in cooperation with DoE has kept 2 mA/cm² as the allowable target fuel crossover current (each for hydrogen and oxygen) in high temperature stable membranes such as the Celtec P® series. This corresponds to a H_2 permeation rate of 7.98 X 10^{-10} mol.cm^{-1}.s^{-1}.bar^{-1} for a 154 μm (or 5.18 X 10^{-10} mol.cm^{-1}.s^{-1}.bar^{-1} for a 100 μm) thick membrane) when hydrogen partial pressure is 2 mbar above atmospheric pressure on the anode side, considered in the course of this work. Similarly, O_2 crossover current of 2 mA/cm² corresponds to 9.97 X 10^{-10} mol.cm^{-1}.s^{-1}.bar^{-1} for a 154 μm (or 6.47 X 10^{-10} mol.cm^{-1}.s^{-1}.bar^{-1} for a 100 μm) thick membrane when oxygen partial pressure is 0.8 mbar above atmospheric pressure on the cathode side (or 4 mbar above atmospheric pressure when air is used on the cathode side of a HT PEM single cell).

2.2 Fundamentals of HT PEM (PBI/H_3PO_4) Fuel Cell operation

Some insights into high temperature PEMFCs containing phosphoric acid doped polybenzimidazole (PBI/H_3PO_4) are enumerated in the current section.

Fuel cells operating above the boiling point of water, offering anhydrous proton conductivity are definitely preferred to classical LT PEMFCs owing to many advantages discussed earlier. Some advantages related to enhanced kinetics and some challenges to high temperature operation are discussed in the following part.

2.2.1 Advantages and disadvantages of higher temperature operation of PEMFCs

At higher fuel cell operating temperatures, reaction kinetics are enhanced commensurate with cell temperature. In PEMFCs, the cathodic current density (i_c) is directly proportional to the cathodic exchange current density (i_{0c}) (or oxygen reduction current density), which determines the rate of reaction at the corresponding cathode catalyst/electrolyte interface. Similarly the anodic current density (i_a) is proportional to the exchange current density for anode reaction (i_{0a}) (or hydrogen oxidation). An increase in temperature in a PEMFC leads to an increase in exchange current density. In a fuel cell, the current density for both cathode and anode are equal, whereas the exchange current density of hydrogen oxidation is much higher (some times some orders of magnitude higher) than that of oxygen reduction. Various groups have reported varied i_{0a} and i_{0c} values corresponding to the specific MEAs (containing highly doped, lowly doped and from different processes) they used in their tests, as shown in the following table 2.4.

Table 2. 4: Exchange current densities for a HT PEMFC (from various authors)
(i_{0a} = anodic exchange current density, i_{0c} = cathodic exchange current density)

Sl No.	Anodic exchange current density i_{0a} (mA/cm²)	Cathodic exchange current density i_{0c} (mA/cm²)	Operating Temp.	Author/Source
1	8.01×10^4	8.01×10^{-4}	170°C	Cheddie and Qingfeng Li, et al. [46,24]
2	2.54×10^6	2.54×10^{-2}	185°C	Cheddie and Savadogo et al. [32,47]
3	3.01×10^5	3.01×10^{-3}	150°C	Cheddie and Savadogo et al. [32,33]
4	-	$2.1 - 4.1 \times 10^{-2}$	130°C to 180°C	Shamardina O and Kulikovsky et al. [48]
5	1.61×10^2	1.61×10^{-6}	150°C	Cheddie and Wang et al. [32,49]

6	1.0×10^4	1.0×10^{-1}	180°C	Broka [50]
7	1.0×10^4	-	180°C	Mamoulk and Scott [51]
8	-	8.3×10^{-1}	180°C	Bandlamudi (this work)
9	-	4.04×10^{-2}	130°C	Bandlamudi (this work)

As can be noticed from the Table 2.4, the anodic exchange current densities are many orders of magnitude higher compared to the cathodic exchange current densities, depending on their specific type. It is known that the lower the exchange current density, the lower the reaction rate and the higher the temperature, the higher the exchange current density for a given system. The lower (i_{oc}) values mean poor cathode kinetics. Therefore the oxygen reduction reaction (ORR) more or less determines reaction kinetics in a PEMFC, hydrogen oxidation being much faster and is not a concern. Further more, increased temperature also means an increase in electronic transfer coefficient "α" for ORR on a Pt electrode. For instance, Song et al [52] had determined the "α_T" (or the transfer coefficient at any temperature "T") to be 0.001678T (T in Kelvin), which is to say that higher temperature operation enhances reaction kinetics in a PEMFC. As for the PBI based membranes used in HT PEMFCs, there are some advantages compared to PFSA (Nafion®) based membranes used in LT PEMFCs. For instance, the electroosmotic drag coefficient in PBI based membranes is close to zero, where as in PFSA based membranes this can be 7 to 8 molecules of water per sulfonic acid site [53] or 2.5 to 3 water molecules per transferred proton, calling for complicated water management in LT PEMFCs. Permeability of gases (such as hydrogen and oxygen) through PBI is lower than in Nafion® based membranes, which implies that a dry PBI membrane serves as a robust barrier between a fuel and oxidant used in a fuel cell. Broka et al [54] had elucidated that in dry Nafion® membranes, at 80°C, permeability of hydrogen and oxygen are in the range of 10^{-11} to 10^{-12} mol.cm^{-1}.s^{-1}.bar^{-1}. Mecerreyes et al [55] had analysed the structure of PBI material and had concluded that PBI polymer membranes are dense with close chain packing with a density of 1.34 g/cm due to its rigid molecular structure and strong hydrogen bonding. Kumbharkar et al [56] had reported hydrogen permeability of 2×10^{-13} mol.cm^{-1}.s^{-1}.bar^{-1} and oxygen permeability of 5×10^{-15} mol.cm^{-1}.s^{-1}.bar^{-1} in PBI membranes at room temperature, which is two to three orders of magnitude lower than in Nafion® based membranes. He et al [57] had reported hydrogen permeability of $1.6 - 4.3 \times 10^{-10}$ mol.cm^{-1}.s^{-1}.bar^{-1} and oxygen permeability of $3 - 9.4 \times 10^{-11}$ mol.cm^{-1}.s^{-1}.bar^{-1} in acid doped PBI membranes in the 80°C – 180°C temperature range. Li et al [58] have elucidated that when PBI membranes are doped with acid (PA), the membranes are swollen resulting in a significant separation of the polymer backbones resulting in an increase of 2 to 3 orders of magnitude in the permeability of hydrogen and oxygen gases when compared to pristine PBI membranes (in the 80°C – 180°C temperature range). Neyerlin et al [59], had determined the hydrogen permeability of acid doped PBI membranes to be in the range of $2 - 2.5 \times 10^{-10}$ mol.cm^{-1}.s^{-1}.bar^{-1}, based on their investigations in a real fuel cell environment. Liu et al [60], had reported lower oxygen permeability rates of 9.6×10^{-13} mol.cm^{-1}.s^{-1}.bar^{-1} based on PBI membrane doped with 6 moles of PA per mole of PBI repeat unit and at 150°C. These values are 2 orders of magnitude lower than those reported by He et al as mentioned above. Whereas in liquid PA (in case of PAFCs), usually the oxygen diffusion coefficient-solubility product best represents oxygen permeability. Klinedinst et al [61] have studied the solubility and diffusivity of oxygen in hot phosphoric acid (100°C – 150°C) of varying concentrations (85 wt% to 100 wt%). They reported about 4.5×10^{-9} mol.cm^{-1}.s^{-1} of oxygen permeability (corrected to 1 bar of oxygen partial pressure on the data they had obtained) in 90 wt% PA at 150°C.

However, there are definitely some challenges to meet, to operate PEMFCs at elevated temperatures (> 100°C). For instance, catalyst oxidation is a known issue related to LT PEMFCs. This is in fact one of the issues to be resolved to ensure long term stability of fuel cells. At higher operating temperatures such as the ones in HT PEMFCs, these problems will be more pronounced. Platinum particle i) agglomeration, ii) dissolution, iii) migration, iv) dissolution of smaller particles and re-deposition of those nano-particles on larger particles, have been some of the known mechanism of catalyst related degradation [62,63,64]. The chemical potential of differently sized Pt particles is different and on the cathode side where the cell has a high operating potential, these processes dominate (resulting in particle growth or agglomeration). Under OCV (high cell potentials) conditions, Pt^{2+} ions coming out of the Pt catalyst can get into the membrane and then redeposit themselves on larger particles leading to Pt particle growth, reducing the total ECSA (Electrochemically active surface area), thus resulting in cell degradation. Some tests performed employing a commercially available MEA under OCV conditions are discussed in chapter 5.1.4. Some groups have been working on realising stable alternatives to existing Pt/C catalysts. Liu et al [65] has developed a Pt/ZrO$_2$/C catalyst and elucidated that it was more stable than a standard Pt/C in a HT PEMFC operating at 150°C.

Besides catalyst degradation, oxidation of catalyst support leading to the loss of supported catalyst is also a challenge in a HT PEMFC environment. Carbon corrosion is a big problem while operating at high temperatures as in HT PEMFCs. The loss of carbon support can also result in decreased hydrophobicity of the carbon surface (as the formation of carbon-oxygen groups makes the surface more hydrophilic).Some attempts have been made by some groups to find alternatives to Vulcan carbon. Wang et al [66] had elucidated that CNTs (carbon nanotubes) have been found to be more corrosion resistant than Vulcan carbon and MWCNTs (multiwalled CNTs) have showed a 30% lower corrosion than Vulcan carbon. Boron doped diamond had exhibited improved oxidation resistance when used as a support for Pt catalysts. Some groups have incorporated a transition metal oxide onto a Pt/C, which offered greater stability. For instance, $Pt/TiO_2/C$, $Pt/ZrO_2/C$, were used as the interaction between metals (Pt and TiO_2) is stronger than between a metal (Pt) and carbon, as reported by Shim et al [67]. Therefore, catalyst and catalyst support related issues still remain as a challenge when PEMFCs are to be operated at elevated temperatures.

Besides membrane, catalyst and catalyst support related issues, oxidant (oxygen) diffusion in acid doped membranes posses yet another challenge. Oxygen has to diffuse through a barrier of PA (electrolyte) in the MEA of a HT PEMFC, before it could reach the reaction sites (or the three phase boundary). Therefore oxygen solubility and dissolution in PA at different operating temperatures dictates the ORR (oxygen reduction reaction), which is also a dominating reaction (compared to HOR or hydrogen oxidation reaction). Furthermore, phosphate anion adsorption impedes ORR, which was observed to be critically dependant on the specific Pt crystal surface as enumerated by He Qinggang, et al [68].

2.2.2 Proton conductivity of acid doped PBI membranes

Li et al [58] had studied proton conductivity in acid doped PBI. He proposed four possible mechanisms resulting in proton conduction in PBI/H_3PO_4 systems as shown in Fig 2.7.

1) Proton hopping from one N (Nitrogen) site to another in a non doped PBI: This mechanism seems to contribute only a little to the conductivity of the membrane. Under humidified conditions, this conductivity was observed to be a little better, as discussed by Hoel [69].

2) Proton hopping from N-H site to a phosphoric acid anion: This mechanism seems to contribute significantly to the proton conductivity of the membrane. For instance, even a doping level of 2 moles of phosphoric acid per mole of PBI repeat unit resulted in a conductivity of

close to 0.01 S/cm at 200°C [70], and these 2 moles of acid are immobilized (bonded to the membrane). The remaining molecules of acid present in the acid doped membrane contribute to proton conductivity.

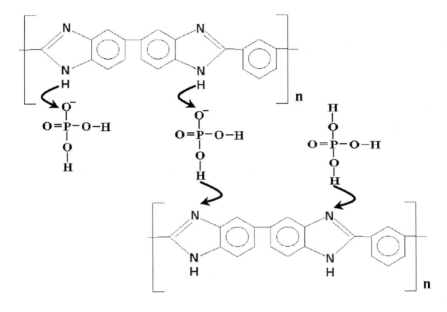

Figure 2.7: Proton conduction mechanism in acid doped PBI (adapted from Li et al)

3) Proton hopping along the $H_2PO_4^-$ anionic chain: This mechanism is possible due to the presence of excess acid or free acid present in the acid doped membrane. Li et al have enumerated that in a membrane doped with 5.7 moles of PA per mole of PBI repeat unit developed by them, they had measured a conductivity of 0.0046 S/cm at room temperature, 0.048 S/cm at 170°C and 0.079 S/cm at 200°C. They had concluded that the presence of free acid contributed to mainly to those proton conductivities. The highly doped membranes from BASF (with about 20 to 30 moles of PA per mole of PBI repeat unit) have conductivities of 0.220 S/cm at 160°C. The presence of excess acid seems to contribute greatly to the proton conductivity of these membranes.

4) Proton hopping via water molecules: The influence of relative humidity on the proton conductivity of acid doped PBI membranes seems to be a minor one. In lowly doped membranes such as the ones developed by Li, et al, the conductivity dependence on relative humidity

was significant only at higher temperatures. For instance, they had reported that an increase of relative humidity from 0.15% to 5% (for a membrane at 200°C), resulted in doubling of conductivity. But the developers of highly doped membranes (such as BASF) contend that their membranes are mostly anhydrous, meaning the proton conductivities of their membranes do not critically depend on water. However, presence of water seems to improve conductivity slightly, depending on the operating parameters (such as load current and cell temperature), even in highly doped membranes.

The anhydrous conductivity of different acid doped PBI systems was studied by Bouchet et al [71]. They had proposed that the proton migrates from one imide site to a neighbouring vacant one. This process is assisted by the counter anion facilitated by the Grotthus mechanism.

3 Cell assembly, Cell and Stack tests

This chapter deals with the experimental investigations performed to characterize high temperature PEMFCs. Cell assembly, cell operation in a test stand, cell impedance measurement using Electrochemical Impedance Spectroscopy (EIS) technique, single cell performance at different operating temperatures such as 180°C, 170°C, 160°C, 150°C, 140°C and 130°C are discussed. The observed open circuit potential of the HT PEMFC at various operating temperatures also is discussed with an appropriate figure.

Carbon monoxide (CO) tolerance of HT PEMFCs is a well known phenomenon. In this work, performance of a single cell when fed with various concentrations of CO (into the anode stream) at cell operating temperatures of 130°C-180°C are discussed. It was interesting to see that a typical HT PEMFC could tolerate even 20% of CO at temperatures close to 180°C, with a loss of about 159 mV at 200 mA/cm² of load current. Whereas, at 130°C with 1% CO in the anode stream, when a load current of 400 mA/cm² was drawn from the cell, the cell had lost most of its voltage and finally the cell voltage had settled at 0.0210 Volts. At the same operating temperature of 130°C, without any CO being fed into the anode, the cell voltage was 0.5551 V at 400 mA/cm². Investigations performed at other cell operating temperatures, CO concentrations, etc are discussed in this chapter. Investigations performed to evaluate the performance of a short stack when fed with 3 types of synthetic reformates is also presented in the following sections of this chapter.

3.1 Single cell characterization

A HT PEMFC single cell was constructed from commercially available cell components as depicted in Figure 3.1. The test stand where the necessary tests were performed is depicted in Fig 3.2. As may be seen from Fig 3.1, the cell hardware consists of (i) a pair of 20 mm thick stainless steel end plates, (ii) one set of isolation layers, (iii) a pair of 10 mm thick bipolar half plates, (iv) one set of high temperature stable elastomer based gaskets, and (v) one MEA (membrane electrode assembly). However, while measuring the impedance of the cell assembly, in some tests, no MEA was used, as explained in section 3.1.1.2.

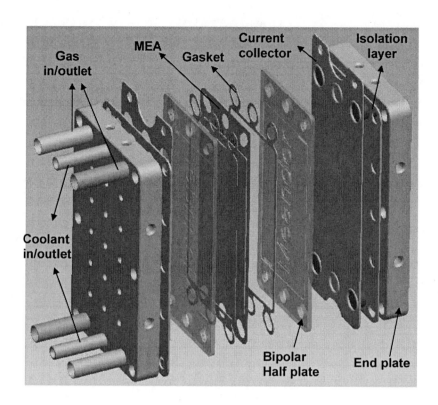

Figure 3.1: Essential components of a HT PEMFC

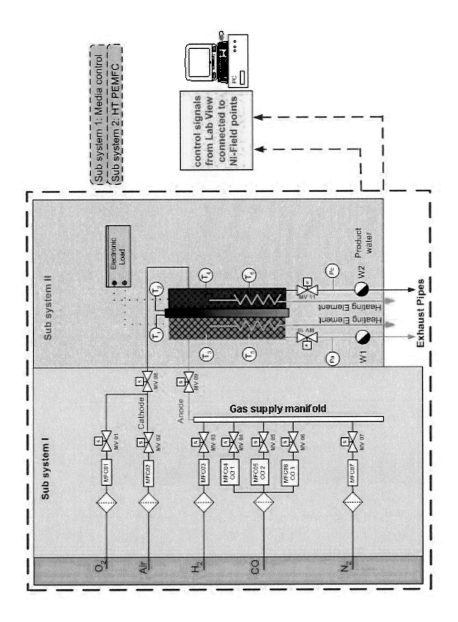

Figure 3.2 : Test stand and component set-up used to characterize HT PEMFC

The 20 mm thick steel plates were chosen to ensure uniform cell component compression even at temperatures close to 200°C. It is known that aluminium, although a good conductor of heat, is a poor choice as end plate material, especially at high temperatures, due to its inferior dimensional integrity when used as a plain plate. However, robust end plates can be constructed based on intelligent design (involving different metal combinations and structural patterns), as enumerated by Evertz et al [72]. The flow field geometry was machined (using an appropriate CNC machine at ZBT) on the plain graphite-polymer based plates purchased from SGL Carbon, Germany. These plates are commercially available as BBP 4 type, which are suitable to be used in a HT PEMFC environment. The perfluor-kautschuk (FFKM) based gaskets were purchased from Precision Polymer Engineering Limited, United Kingdom. The Celtec P 2000 MEA was purchased from BASF fuel cells. The current collector plates used were based on 1 mm thick copper plates coated with 1 μm thick Gold layer with an intermediate layer of Nickel (20 μm thick) (separating the copper and the Gold layers). Isolation layer was made from hard PTFE sheets of 1 mm thickness. Six screws (M6 type), each coupled with a pair of flat washers and 6 spring washers were used to hold all the cell components together. The gas in and outlet pipes were made of stainless steel and PTFE pipes that fit into these steel pipes (see Fig 3.3). This type of a connecting pipe ensures that only PTFE pipe comes into contact with the bipolar half plates whereas the stainless steel housing can be easily connected to the standard metal connector that runs between a MFC (mass flow controller) and the fuel cell hardware, as will be explained later in the current chapter, while discussing the test stand set up and the diagnostic tools used.

Figure 3.3: PTFE lining inside a stainless steel housing used in the HTPEMFC

A single cell used for further tests was assembled with all the mentioned necessary components and was fixed in the test station depicted in Fig 3.2.

Test stand layout and hardware to control and monitor HT PEMFC

A test stand was constructed with mass flow controllers (MFCs) for oxygen, air, hydrogen, carbon monoxide and nitrogen, as depicted in Fig 3.2. The subsystem I in the test stand layout shows the way these MFCs (MFC 01 to MFC 07) are connected to the HT PEMFC. Each MFC has a solenoid valve connected to it. When any MFC is not used, the connected valves were automatically closed, thus not allowing any back flow into MFCs from the gas supply manifold (long connecting pipe) shown in Fig 3.2. The gas supply (mains) points are shown in the yellow block (sub system I), which are connected to pressure regulators (diamond blocks) that are connected to respective MFCs. Two gas supply manifolds (one for anode and the other for cathode represented by red and blue lines respectively) are connected to the HT PEMFC via safety valves (MV 08 and MV 09). These valves ensure that the inlets of the fuel cell are closed when the cell was not in operation, thus safe guarding the electrolyte from being affected by the connected pipes and the media in contact. Similarly the valves (MV 10 and MV 11) were fixed to the outlet (anode and cathode respectively) of the fuel cell. W1 and W2 are water collecting bottles connected to the outlet streams of the anode and cathode. The entire HT PEMFC cell hardware with these water traps are located in subsystem II of the test stand layout shown. An electronic load (EL 3000A) from Electro-Automatik GmbH, Viersen, Germany, which has the possibility to measure voltages close to 0 Volts was connected to the fuel cell as depicted in the test stand layout. However, electronic load from Basytec ® coupled to Gamry's EIS apparatus was connected to the HT PEMFC, in place of this electronic load, while performing impedance measurements. The outlets of the fuel cell are connected to the gas exhaust pipes installed in the laboratory. Heater cartridges were inserted into the end plates of the fuel cell and their terminals were connected to the electricity supply mains in the laboratory. The entire test equipment was monitored and controlled by a computer loaded with Lab View ® software from National Instruments and the field points connected to each hardware element. They include temperature elements, solenoid valves, electronic load, MFCs, pressure sensors. The HT PEMFC's cell voltage was monitored by sensors connected to the bipolar half plates of the fuel cell. Various protocols were implemented for start up, shut down and operation of the fuel cell depending on the tests performed. For instance, while feeding the fuel and oxidant to the fuel cell, care was taken not to exceed a differential pressure of ~ 300 mbar (across the MEA). The safety protocol implemented in Lab View® ensured that in the event of any excess differential pressure across the fuel cell MEA, the gas supply was cut off, load was disconnected, the in and outlets of the fuel cell were closed and about 1 litre of N_2 gas was allowed into the anode of the fuel cell (to bring down the cell voltage level from its open circuit potential). In case of any gas leaks, the sensors installed in the test stand (very close to the fuel cell) would send an electrical signal, thus implementing a safety protocol. The outlets of the anode and cathode sides of the HT PEMFC were directed through a two stage condenser system where most of the water was collected. Water was removed from these bottles and was transferred into other clean bottles which were sufficiently rinsed with de-ionized water and were labelled accordingly, before they were analysed to find out elements present in the water.

3.1.1 Ohmic overvoltage

HT PEMFC's performance loss due to ohmic overvoltage was investigated, with an aim to understand impedance contribution from various cell components at different temperatures (up to 180°C). To achieve this, tests were performed and the following resistances were measured using EIS (electrochemical impedance spectroscopy) device.

a) Cell resistance versus cell compression (cell assembly + MEA + cable) (Fig 3.8)

b) Cell resistance versus cell temperature (cell assembly without MEA + 1GDL+ cable) called in this work as (cell assembly - scenario I) (Fig. 3.9)

c) Cell resistance versus cell temperature (cell assembly + MEA + cable) called in this work as (cell assembly - scenario II) (Fig. 3.12)

d) Resistance of cell components from 4-pole conductivity measurement device (Table 3.3)

e) Cell resistance versus cell temperature (cell assembly without MEA+2GDL+2CL+cable)

$$= b) + 1 \ GDL + 2 \ CL \ (Fig \ 3.14)$$

f) Membrane resistance = c) – e) (Fig 3.16)

In the case of a), b) and c) resistance measurements, cable connecting the HT PEMFC cell assembly with the EIS terminals was included. In the case of d), a separate test was performed using the 4-pole conductivity measurement device (Fig 3.13) connected to EIS device from Zahner Messtechnik (IM6). In this case, resistance of the bipolar half plates, current collector plates and the GDL were studied, piece by piece (Table 3.3). Cell resistance in case of e) was computed from the resistances measured in case of b), d) and the catalyst layer conductivities were obtained from literature [73]. Membrane resistance in case of f) was obtained by simply subtracting e) from c). Membrane's conductivity was calculated and all the measured resistances are depicted in Fig. 3.12. The total resistance contributions from all the components used in the HT PEM single cell are shown in Table 3.4.

A HT PEMFC's ohmic overvoltage is the cell voltage loss caused due to the total cell resistance which is specific to a) cell temperature, b) compacting force exerted on the cell components, c) type of the components used. The cell temperature would influence mainly the membrane's conductivity; the compacting force would influence the contact resistance between cell components such as CC/BPP, BPP/GDL, GDL/CL.

Furthermore, the type of components used such as the gold plated copper current collector, graphite compound based bipolar half plates, the materials used in the GDLs, the polymer binder contained in the catalyst layer, membrane material and its thickness, all of them will have their respective influence on cell resistance. Cell resistance is the high frequency impedance at which the imaginary impedance is zero. Total cell resistance is the sum of resistance contributions from all the cell components. Also, total cell impedance is the sum of resistive as well as complex (inductive or capacitive or a combination of both) parts, which follow the relation shown in Eq 33.

Total cell impedance, $Z = R + j X$, where	
	Eq. 33
'R' represents total resistive part of cell impedance and	
'X' represents total imaginary impedance in an electrochemical cell	

However, as can be clearly seen from Table 3.4, the resistance contribution from the membrane is usually the highest, followed by bipolar plates, current collector plates, GDLs and the catalyst layers. As the bipolar half plates used are based on graphite-polymer compounds (also, each plate was about 10 mm thick), their contribution to the total cell resistance also can not be ignored. The bipolar plates used in this test are the high temperature stable BBP4 type from SGL Carbon Company based in Germany. If metallic bipolar plates were used, instead of these graphite based ones, the contribution to cell resistance could have been very low (provided the passivation layers coated on them do not exhibit higher contact resistances). However, chemical stability of metallic bipolar plates is still a subject of research.

As for the complex part (in Eq.33), the double layer capacitance at the electrode membrane interface, inductances of leads present in the entire circuitry of the fuel cell, contribute to the overall cell impedance. The total high frequency impedance (or ohmic resistance) of the HT PEM single cell used in this study is depicted in Fig 3.15. However, the total impedance of the cell dictates the total loss in cell performance (or loss in cell voltage) in the ohmic region of the cell performance plot shown in Fig 2.2. Furthermore, the total heat produced in a fuel cell mostly comes from this total impedance of the cell. As a fuel cell in most cases is expected to generate electricity and not heat, the impedance of the cell has to be kept low, to avoid any production of unwanted heat. Having said that, the heat produced in a HT PEMFC cell or stack will help maintain it at the required operating temperatures of 160°C – 180°C.

Electrochemical impedance spectroscopy (EIS) is a technique which involves the application of an a.c. (alternating current) stimulus, either in the form of current or voltage and measuring the response. In potentiostatic mode, an a.c. voltage (excitation) is applied across the electrochemical cell and the cell's response can be measured as its impedance at different frequencies. In galvanostatic mode, an a.c. current (excitation) is sent through the electrochemical cell and the cell's response is measured as its impedance at different frequencies. The impedance spectrum (impedance plot at different frequencies) is interpreted in the form of equivalent circuits, depending on the system contained in the electrochemical cell considered [74]. The high frequency impedance (at frequencies >1 kHz) at which the imaginary part (second term in Eq.33) is zero, is taken as cell resistance (in Fig 3.4, it is shown as R_{el}). At high frequencies, the impedance contribution from capacitive and inductive processes is usually very low, whereas at low frequencies (< 1 Hz), major contribution comes from the inductance and capacitance of the electrochemical system (in this case fuel cell). Typically an EIS spectrum can be drawn as shown in Fig 3.4, X-axis representing the real part of the impedance shown in

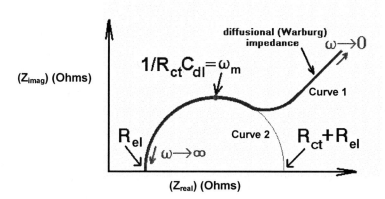

Figure 3.4: Nyquist plot of Randles model. Curve 1: Mixed charge transfer and diffusion controlled EIS spectrum; Curve 2: Simplified Randles model with only charge transfer control.

Eq.33 and its imaginary part on the Y-axis [75]. A plot of real part versus imaginary (or complex) part of the EIS spectrum is termed as nyquist diagram of cell impedance. R_{el} in Fig 3.4 represents resistance contribution from the fuel cell membrane. However, the total cell resistance (both ionic and electronic) from all the cell components is represented by R_{el} as discussed earlier. Also, as can be seen from Fig 3.4, R_{el} is the high frequency (ω tending to infinity) intercept of the nyquist plot on its x-axis at which the imaginary impedance contribution is zero. The radial frequency $\omega = 1/2\pi f$, where f stands for frequency in Hertz.

Curve 1 in Fig 3.4 represents Randles model as it was proposed by J.E.B.Randles in 1947. This Curve 1 represents mixed kinetic and diffusion process at low frequencies (ω tending to zero). Curve 2 represents simplified Randles model which is in the form of a semi-circle with its low frequency x-axis intercept representing ($R_{el} + R_{ct}$) where R_{ct} stands fro charge transfer resistance. Also, the frequency (ω_m) at which the ($-Z_{imag}$) value is the maximum equals ($1/R_{ct}C_{dl}$) where C_{dl} stands for double layer capacitance. As R_{ct} can be obtained from the low frequency intercept of the nyquist plot as shown in Fig 3.4, in the case of a simplified Randles model, C_{dl} can be calculated from the known values of ω_m and R_{ct}. In the case of Curve 1, where the diffusion process can be semi-infinite or infinite, the low frequency x-axis intercept tends to be infinite. The double layer capacitance C_{dl} comes from the electrical double-layer that is formed at the electrode/electrolyte interface in a fuel cell. The charge transfer resistance R_{ct} represents the resistance associated with the charge transfer processes involved in fuel cell reactions.

Usually equivalent circuits can be modelled to evaluate parameters such as R_{ct}, C_{dl}, R_{el} as depicted in Fig 3.5 and Fig 3.6. Fig 3.5 represents the equivalent circuit for Randles model with mixed charge transfer and diffusion processes discussed earlier using Curve 1 in Fig 3.4. Fig 3.6 represents the equivalent circuit for simplified Randles model with only charge transfer process discussed earlier using Curve 2 in Fig 3.4.

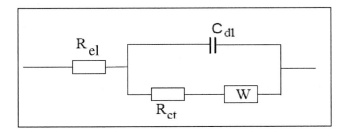

Figure 3.5: Equivalent circuit for Randles model. R_{el} represents ohmic resistance, C_{dl} represents double layer capacitance, R_{ct} stands for charge transfer resistance and W stands for Warburg impedance.

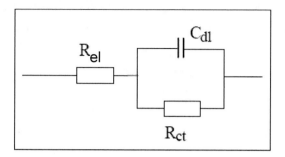

Figure 3.6: Equivalent circuit for simplified Randles model. R_{el} represents ohmic resistance, C_{dl} represents double layer capacitance, R_{ct} stands for charge transfer resistance.

The Warburg impedance (W) represents diffusion of species into the electrode (Fig 3.5). The Warburg impedance [76] is a function of Warburg element (WE) which in turn is a function of the diffusivity of the species involved (such as hydrogen, oxygen in the case of a fuel cell reaction), the rate of reaction at the electrodes, concentration of the species, current density and the cell potential. There is no simple element to model the Warburg impedance. Also, various combinations of purely resistive, inductive or capacitive behaviours do not always describe the electrochemical processes involved while acquiring the EIS spectrum. Therefore, distributed elements are used to model such a complex behaviour. These distributed elements can be represented in a model circuit as constant phase element (CPE) or Warburg element. The impedance of CPEs [77,78,79,80,81,82] are attributed to (i) the double layer charging behaviour when the electrode surfaces are rough and (ii) non-ideal double layer behaviour, ideal behaviour being in a pure capacitor. The impedance of a CPE is represented as Eq.34.

$$Z_{CPE} \text{ (A/cm}^2\text{)} = \left[\frac{1}{T*(j\omega)^{\varphi}} \right] \qquad \text{Eq.34}$$

T and φ are CPE parameters in Eq.34. The radial frequency ($\omega = 1/2\pi f$), where f is frequency in Hertz. Also, when $\varphi = 1$, the CPE parameter T represents (C) or pure capacitive behaviour. When $\varphi = -1$, the CPE parameter T represents (1/L) or pure inductive behaviour. When $\varphi = 0$, the CPE parameter T represents (1/R) or purely resistive behaviour. When $\varphi = 0.5$, the CPE parameter T represents the Warburg element (WE) and the impedance spectrum obtained from a EIS scan typically takes the shape shown in Fig 3.7. "f" in Fig 3.7 stands for the scanning frequency, Z_{real} stands for the real part, Z_{imag} for imaginary part and $-Z_{imag}$ (max) stands for the maximum negative imaginary part. Also, Fig 3.7 represents semi-infinite mass diffusion into porous electrodes with its 45° angle behaviour at higher scanning frequencies. This model was used for tests performed and discussed in 3.1.2.

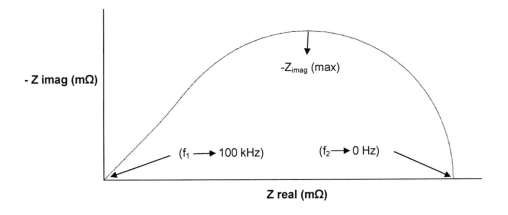

Figure 3.7: Nyquist plot for Randles model when φ (CPE parameter) = 0.5. f is frequency, Z_{real} stands for real part of the impedance and Z_{imag} for imaginary part. ($-Z_{imag}$) max stands for the maximum negative imaginary part of cell impedance.

In the tests performed (tests a – d) explained further, after applying 10 mV (a.c.) across the HT PEMFC single cell, the impedance of the cell was measured in potentiostatic mode of the EIS. The high frequency impedance (1 KHz – 3 kHz) at which the inductive and capacitive impedance contributions are zero or (the cell resistance) was noted down, as enumerated in the following sections.

3.1.1.1 Cell Impedance at different levels of cell compression (test a)

Investigations were performed to understand the HT PEM fuel cell's internal impedance at different cell compressions. It is known that lower cell resistance (or high frequency impedance) is desired to ensure lower ohmic overvoltage in a fuel cell. But usually this could be achieved only after ensuring adequate compressive force on the end plates of the fuel cell. Too little a compression would result in large contact resistances and too much compression might be detrimental to the gel based membrane employed in the cell. Therefore, it is essential to identify optimum cell compression that would minimize contact resistances, keep the membrane intact and offer minimum possible cell impedance. The single cell was compressed with six M6 screws, fixed to the stainless steel end plates. Initially the cell compression was only hand tight. Then, using a spanner with pre-defined torque, the compressive force was increased to about 0.75 N/mm² (or 7.5 bar) in about 4 steps. Then, the anode (negative) and cathode (positive) terminals of the cell were connected to an EIS analyzer (Electrochemical Impedance Spectroscopy) from Zahner Messtechnik, to measure the impedance of the cell. The inlets and outlets (anode and cathode sides) of the fuel cell were closed initially, and nitrogen gas (N_2) from gas mains in the test stand was allowed into the cell for about a minute and then, the inlets and outlets were closed. After applying 10 mV (a.c.) across the cell, the impedance of the cell was measured in potentiostatic mode of the EIS. The high frequency impedance (1 KHz – 3 kHz) at which the inductive and capacitive impedance contributions are zero or (the cell resistance) was noted down. In a similar manner, tests were performed up to a cell compressive force of 2.8 N/mm² or 28 bar, at regular intervals. The measured cell resistance as a function of cell compressive force is shown in Fig 3.8.

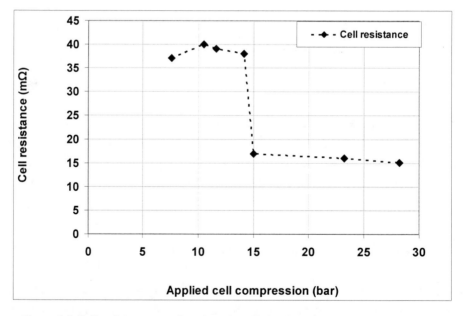

Figure 3.8: Cell resistance as a function of applied cell compression

As can be seen from Fig 3.8, the resistance of the cell assembly was initially 37 mΩ at a cell compression of 0.75 N/mm². After increasing the cell compression to 10.5 bar, the cell impedance went up to about 40 mΩ. From then on, the cell impedance began to fall with a corresponding increase in cell compression. It might not be intuitive to see an initial increase in cell impedance, but it may be noted here that the MEA employed here has a highly doped membrane (PBI doped with H_3PO_4). And with a hand tight compression applied on the cell assembly, not much acid was squeezed out of the MEA, initially, therefore the observed lower cell impedance at start. But as the compression reached about 7.5 bar, then, part of the excess acid which was present in the MEA had exited, possibly giving rise to an increase in cell resistance. However, the contact resistances of the cell, namely, the BPP/GDL, BPP/CC, GDL/CL interfacial resistances would come down as the cell compression increases. The dimensional change of the membrane and the subsequent gain in ionic conductivity (due to lower ionic resistance) is not a dominant phenomenon caused due to increased compressive force. As can be seen from Fig 3.8 till 14 bar, there was no substantial fall in cell impedance. Whereas, at 15 bar, the cell impedance fell from 37 mΩ to 16.9 mΩ very rapidly.

From this test, it was observed that the reduction in various contact resistances of the cell, with an increase in cell compressive force up to 15 bar has resulted in a substantial fall in cell impedance. After this point, the fall in cell impedance was not much even with an applied compressive force of about 28 bar. It may be noted here that these investigations were performed at room temperature and pressure (RTP). It may be concluded (see Fig 3.8) that any reasonable cell impedance level could be achieved only after ensuring a cell compressive force of about 15 N/mm². With an increase in cell temperature, the impedance of the cell would come down further, as discussed in the following section.

3.1.1.2 Cell resistance at different temperatures (test b)

Firstly, a single cell was assembled with one set of current collectors, a set of bipolar half plates, isolation layers and the end plates. In this assembly, there was no MEA used, instead, a GDL was sandwiched between the bipolar half plates. The entire cell assembly with the said GDL was compressed to about 2 N/mm² (or 20 bar of compressive force) at RTP. Then, after connecting the current collector terminals to the EIS (Zahner Messtechnik) measurement device, impedance (1 KHz – 3 kHz) of the cell assembly was measured at different temperatures. The cell assembly was kept at the required temperature by heating the electrical heaters inserted in the end plates of the cell assembly. Keeping the EIS device in potentiostatic mode after applying a signal of 10 mV (a.c), resistance of the cell assembly was recorded at different temperatures (1 kHz – 3 kHz) and is depicted in Fig 3.9.

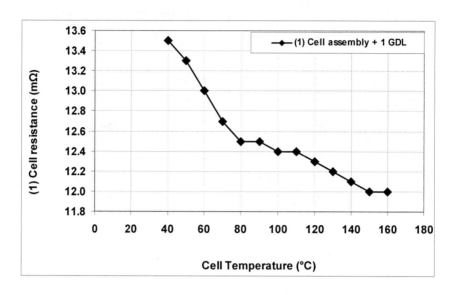

Figure 3.9: Resistance of the cell assembly versus cell temperature (without MEA)

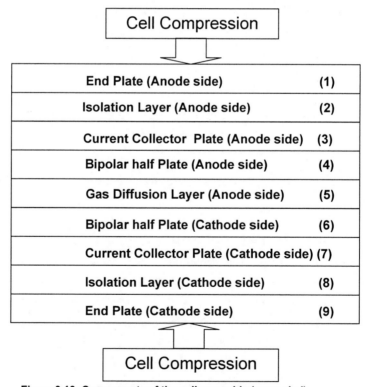

Figure 3.10: Components of the cell assembly (scenario I)

Table 3.1: Components contributing to resistance of cell assembly in Fig 3.9

SI No.	Component	Thickness (mm)	Quantity
1	BPP	10	2
2	Current collector	1	2
3	GDL	0.300	1

The components that were used in this cell assembly (corresponding to measurements shown in Fig 3.9) are shown in Fig 3.10. Note that the HT PEM fuel cell and the cables used to connect the cell to the EIS device (termed here as cell assembly – Scenario I), contributed to the measured resistance shown in Fig 3.9.

It has been observed that, from the cell assembly (without MEA) depicted in Fig 3.10 and Fig 3.9, major contribution to cell resistance was from graphite compound based bipolar plates, whereas the resistance contribution from the current collector plates and from GDL was almost negligible, as expected. The resistance contribution from the connecting cables also was not prominent. In the following sections, impedance contributions from these cell components will be discussed further. The current investigation without a MEA, but with a GDL was performed to understand the impedance of cell assembly, contributions from BPP/GDL and BPP/CC contact layers at different temperatures of the cell assembly. Although the resistance of the cell assembly shown in Fig 3.9 is rather high (even without a MEA), this test was performed to compare the cell resistance with a MEA under identical test conditions and then to ascertain the membrane conductivity. The observed change in cell resistance from 13.5 mΩ to 12 mΩ (Fig 3.9) comes from the changes in electrical conductivity of cell components as the cell temperature was raised to 160°C at regular intervals.

A second test was performed (**test c**), this time, with a MEA inside the cell assembly. This second test had the cell component layout as shown in Fig 3.11. The only difference between the two cell assemblies, namely, cell assembly of (Fig 3.11) and the one in (Fig 3.10) is the MEA in the former, which includes 2 Gas diffusion layers, 2 catalyst layers and one membrane. Like in the first case, after compressing the cell assembly to 2 N/mm² or 20 bar, impedance measurements were performed with EIS (from ZahnerMesstechnik, Germany) device, in the same test stand, under the very same test conditions. The connecting cables, and the test set up was identical to the first set up explained earlier. The measured cell resistance is plotted against cell temperature and is shown in Fig 3.12. Contribution to the cell assembly resistance shown in Fig 3.12 comes from the components shown in Table 3.2. It can be seen from Fig 3.9 and Fig 3.12, that the resistance contribution from the membrane is rather high. From these measured values of membrane resistance, membrane conductivity at different temperatures was calculated. To arrive at this, some more tests were performed to understand the resistance contribution from other cell components.

Thirdly, the impedance of bipolar half plates, GDL and the current collector plates were measured separately (**test d**) using a 4-pole impedance measurement device at ZBT which is shown in Fig 3.13. Resistance of the catalyst layer is taken from literature [73]. The measured values of various component resistances are shown in Table 3.3. These resistance values refer to a compression of 2 N/mm², which has been adopted as a standard in the current work.

Now, to evaluate the resistance contribution from the PBI/H_3PO_4 membrane, the measured resistance of the cell assembly with one GDL (Figure 3.9) based on resistance of the cell assembly shown in Fig 3.10, plus the measured resistance of one more GDE (Table 3.3) plus resistance of one more catalyst layer was computed (**test e**) and is shown using the plot (cell resistance versus operating temperature) in Fig 3.14. As the high frequency impedance contribution from the catalyst layers and GDL are not significant, no appreciable change can be observed when compared to the first case, where only one GDL was used (refer to Fig 3.9).

Now, the difference between the cell resistances shown in Fig 3.12 (with MEA) and in Fig 3.14 (without MEA, but with two GDLs and two catalyst layers) represents the membrane resistance (**test f**) at different cell operating temperatures and is shown in Fig 3.15. The observed thickness of the membrane was around 152 μm and the area of the membrane used was 50 cm². Conductivity of the membrane was calculated using these values and is shown in Fig 3.16.

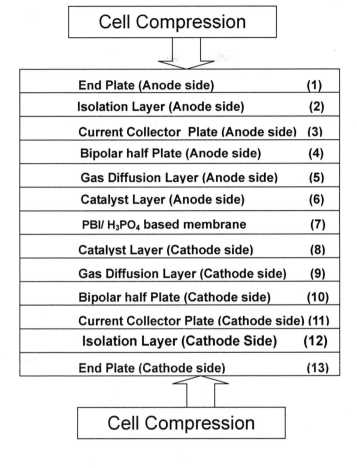

Figure 3.11: Components of the cell assembly (scenario II)

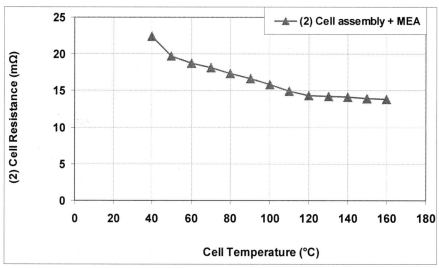

Figure 3.12: Resistance of cell assembly versus cell temperature (with MEA)

Table 3.2: Components contributing to resistance of cell assembly shown in Fig 3.12

SI No.	Component	Thickness (mm)	Quantity
1	BPP	10	2
2	Current collector	1	2
3	GDL+CL (or GDE)	0.36	2
4	Membrane	0.152	1

Table 3.3: Resistance of cell components from 4-pole measurement device

SI No.	Component	Thickness (mm)	Resistance (mΩ)
1	BPP	10	0.422030
2	Current collector	1	0.015038
3	GDL+CL (or GDE)	0.36	0.000655

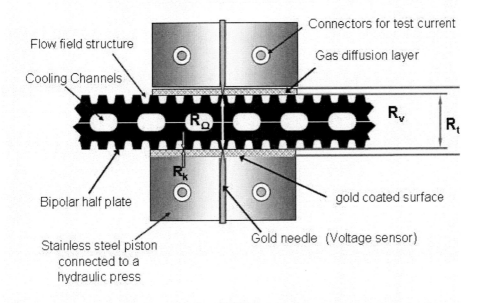

Figure 3.13: Four-pole impedance measurement device set-up

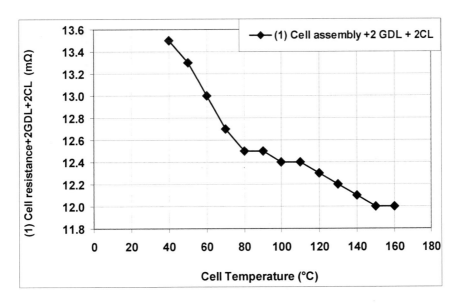

Figure 3.14: Resistance of cell assembly as a function of operating temperature

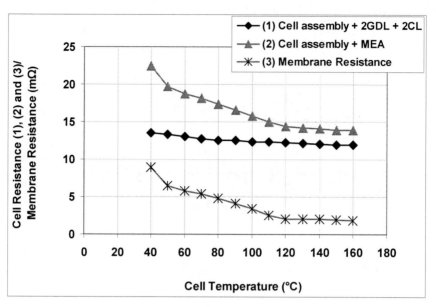

Figure 3.15: Resistance of cell assembly (with and without MEA) and of the
membrane as a function of cell operating temperature

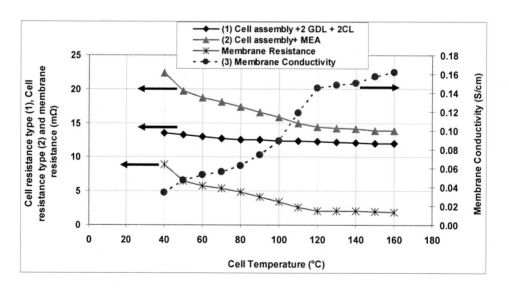

Figure 3.16: Resistance of cell assembly (1, 2 and 3) and conductivity of the membrane
as a function of operating temperature

The membrane resistance was calculated by deducting the values of curve (1) shown in Fig
3.16 from those of curve (2). Curve (3) represents the conductivity for a 152 µm thick mem-

brane with an area of 50 cm². The membrane conductivity can also be expressed using the standard Arrhenius form shown in Eq 35.

$$\sigma_{membrane} \ (S/cm) = \left[\frac{\sigma 0}{T} \right] * e^{(-\Delta Ea/RT)} \qquad Eq.35$$

Where

σ_0 = pre-exponential factor for membrane conductivity

E_a = Activation energy (J/mol)

R and T have their usual meaning

The measured conductivities are plotted as Ln ($\sigma \cdot$T) (logarithm of conductivity-temperature product) versus the reciprocal of temperature (1000/T) shown in Figure 3.17.

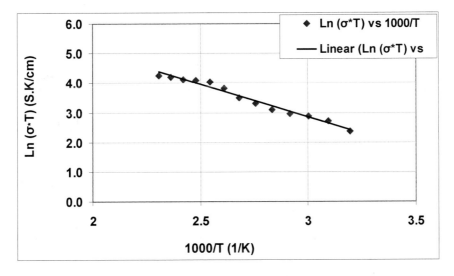

Figure 3.17: Membrane conductivity in the Arrhenius form: Ln (σ*T) vs reciprocal of temperature

The Arrhenius equation for membrane conductivity can be also written as shown in Eq. 36. From Figure 3.17, the relation of the form shown in Eq. 36 was deduced and is shown in Eq. 37.

Ln (σ*T) (S.K/cm) = Ln (σ0) - ΔEa/RT	Eq.36

Ln (σ*T) (S.K/cm) = 9.5081 - 2.2233*(1000/T)	Eq.37

Therefore, from the measured values of membrane conductivity (Fig 3.16), the pre-exponential factor σ_0 was found to be 13466 and the activation energy (Ea) was found to be 18484 J/mol.

It may be seen from Figure 3.16 that the resistance of the cell assembly (without a MEA) shown in curve (1) is more or less constant from 40°C to 160°C, whereas the one with a MEA or curve (2) showed a rapid fall in cell resistance with increasing temperature. Also, after 140°C, the fall in cell resistance is not so significant. It is therefore imperative that a typical HT PEMFC's recommended operating temperature is above 140°C from the point of view of internal ohmic losses. The membrane resistance value for this membrane at 140°C was 2.02 mΩ, at 150°C, it was 1.93 mΩ and at 160°C, it was 1.88 mΩ. The resistance of various cell components, which were measured as detailed above are summarized in Table 3.4.

Table 3.4: Resistance of cell components including the membrane (HT PEM Single Cell)

SI No.	Component	Quantity	Resistance (mΩ)
1	Bipolar half plates	2	0.844060
2	Current collector	2	0.030075
3	GDL+CC (or GDE)	2	0.001309
4	Membrane (152 μm) at 160°C	1	1.880000
	Total cell resistance		2.755444

The total resistance of 2.75 mΩ shown in Table 3.4, represents the total resistive contribution from all the cell components of a HT PEMFC operating at 160°C. This amounts to a high frequency impedance of about 137 mΩ-cm² (area specific resistance), for a 50 cm² cell. Under the mentioned cell compression force of about 20 bar over the entire active area of the HT PEMFC, the contribution from BPP/GDL and BPP/CC were not prominent.

In conclusion, the membrane resistance, total cell resistance (membrane resistance + resistance contribution from rest of the cell components) and the corresponding voltage loss in a

HT PEMFC single cell (with Celtec P 2000 MEA) at different load currents of 20 mA/cm² to 400 mA/cm² at temperatures of 130°C to 180°C are computed and are summarized in Table 3.5. The membrane's resistance at no load (0 amps) but at different temperatures is shown in the third column in Table 3.5. The total electronic resistance of all the remaining cell components according to Table 3.4 is 0.000875 Ω. The total cell resistance is the sum of ionic and electronic cell resistances shown in the fourth column in Table 3.5. The cell voltage loss caused due to the total cell resistance at different load currents is calculated and shown in columns 5 – 9 in Table 3.5.

Table 3.5: Cell voltage loss caused due to ohmic losses (HT PEM Single Cell)

Temp (°C)	Temp (°K)	Membrane resistance (Ω)	Total cell resistance (Ω)	Single cell voltage loss due to cell resistance at different load current densities (Volts)				
				20 mA/cm²	100 mA/cm²	200 mA/cm²	300 mA/cm²	400 mA/cm²
130	403.15	0.002047	0.002922	0.002922	0.014611	0.029221	0.043832	0.058442
140	413.15	0.002024	0.002899	0.002899	0.014495	0.028990	0.043485	0.057979
150	423.15	0.001929	0.002804	0.002804	0.014021	0.028042	0.042063	0.056084
160	433.15	0.001877	0.002752	0.002752	0.013760	0.027520	0.041281	0.055041
170	443.15	0.001830	0.002705	0.002705	0.013526	0.027051	0.040577	0.054103
180	453.15	0.001775	0.002651	0.002651	0.013253	0.026507	0.039760	0.053013

3.1.2 Changes in membrane resistance with hydration in HT PEMFCs

It is well known that classical LT (low temperature) PEMFCs depend on the presence of water for achieving adequate proton conduction. But membranes used in HT PEMFCs are expected to be anhydrous, where proton conduction in theory is achieved via electrolytes that conduct protons, offering adequate proton conductivity (of about > 0.1 S/cm) even at temperatures of 130°C-200°C, without the aid of water. However, presence of water does play a role in these membranes, although they are essentially anhydrous. Some brief introduction to anhydrous proton conduction in typical HT PEMFCs was discussed earlier in this work. But in the current section, commercially available membranes and their properties (especially their anhydrous behaviour) are examined with experiments.

Although there exist quite a number of high temperature stable proton conducting membranes, much attention has been given to PBI/H_3PO_4 based membranes as they have the potential to offer high proton conductivity. The phosphoric acid doped PBI membranes, initially developed by Wainright et al [29,83,84] had offered little ionic conductivity in the early 1990s, however, rapid progress has been made in recent years (up to 2009) in the area of high temperature stable membranes, especially PBI/H_3PO_4 types. Proton conductivity of these membranes, differ vastly, depending on their doping process. For instance, Xiao and Benecewicz [85] have doped PBI (base material) with phosphoric acid by sol-gel process in which PPA (poly phosphoric acid) was hydrolysed to PA (phosphoric acid) and the membrane was finally prepared after a sol-gel transition of PPA and PBI. These membranes had exhibited very high proton conductivites (and very high doping of 32 to 42 mole of PA per mole of PBI repeat unit), which later on were available as BASF membranes. Whereas, PBI was doped with PA by dissolving them with DMAc (Dimethylacetamid) by Li et al [86,87]. They found that only 6.6 to 10 moles of PA per mole of PBI repeat unit could be doped in their process. Obviously proton conductivity of this lowly doped material was lower. Savinell et al had doped PBI with PA by mixing them in TFA (Trifluoroacetic acid) and found the conductivity of the resulting membranes to be lower. There exist other high temperature stable membranes such as AB-PBI/ H_3PO_4 based membranes from FumaTech, Germany, acid doped TPS based membranes from Advent technologies, Greece. Elcomax (formerly Sartorius), Germany also has developed their own variety of PBI/ H_3PO_4 based membranes. But as the acid doping level per PBI repeat unit, to date is the highest for the membranes developed by BASF, their conductivity is the highest. These membranes can be broadly categorized as shown in Fig 3.18.

3.1.2.1 Lowly doped and highly doped high temperature PEMFC membranes

Membranes with PA/PBI doping levels of more than 20 are considered as highly doped and those with PA/PBI doping levels of less than 10 are considered as lowly doped membranes in this work. Various lowly doped and highly doped membranes (as MEAs in HT PEMFCs) have been investigated within this work:

- A single cell with a highly doped membrane was tested in the 0 – 400 mA/cm² load current region with H_2 and O_2 as well as with H_2 and Air to study its EIS spectra with a particular focus on the development of its membrane resistance.

- The same single cell with a lowly doped membrane in the 0 – 400 mA/cm² load current region at 160°C, 180°C and 190°C was operated to study the EIS spectra.

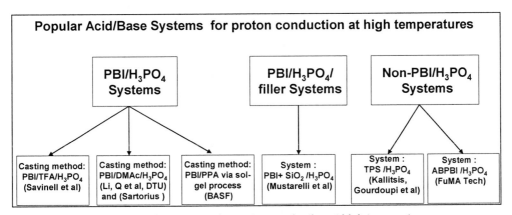

Figure 3.18 : Popular acid/base systems for proton conduction at high temperatures

A comparison is being made here, between a lowly doped and a highly doped system. This analysis is based on impedance spectra (EIS measurements) taken from single cell tests, and focuses on the changes in membrane (electrolyte) conductivity with hydration. As is obvious, higher acid content (PA/PBI repeat unit) results in higher proton conductivity. Li et al have elucidated that 2 moles of PA per every mole of PBI repeat unit is immobilized and the rest of PA present in the doped membrane which usually is referred to as (X-2) moles of PA (or excess acid), contributes to proton conductivity in a typical PA doped PBI based membranes. The notation X stands for the total doping level. Table 3.6 illustrates the polymer/acid contents both as mole ratios as well as weight % pertinent to low and highly doped systems.

Table 3.6: Membrane contents in lowly doped and highly doped PBI/PA membranes

System type	Membrane Content	moles	weight%
Highly doped system	PBI	1	7%
	H3PO4	40	93%
Lowly doped system	PBI	1	31%
	H3PO4	7	69%

The following tests were performed to study the dependence of membrane's impedance on its hydration. Firstly, a highly doped membrane (BASF's Celtec P® 1000 MEA) containing HT PEMFC consisting of the very same cell assembly mentioned in the earlier sections was connected to EIS device from Gamry® coupled to an electronic load from Basytec®. After switching to galvanostatic mode of the EIS, by keeping the perturbation signal as 0.1 A (a.c.), at scanning frequencies of 100 mHz to 100 kHz, at load currents of 0 to 400 mA/cm², the impedance spectra were obtained. The HT PEMFC was maintained at 170°C and was fed with fuel H_2 and oxidant as air with stoichiometries of 1.35 and 2.5 respectively and H_2 and O_2 as fuel and oxidant at stoichiometries of 1.35 and 9 respectively. The data logged is depicted in the form of a nyquist plot (real versus imaginary part of the cell impedance) which is shown in Fig 3.19.

The high frequency intercept (with zero imaginary impedance) of the nyquist plot represents mostly electrolyte resistance (R_{el}) of the cell used as discussed in vivid detail in section 3.1.1. The electrochemical behaviour of the electrode/electrolyte system studied here is analogous to a Randles model discussed previously using Fig 3.5 and Fig 3.7. The low frequency impedance could mainly be attributed to double layer capacitance (C_{dl}), charge transfer resistance (R_{ct}) and Warburg element W. The Warburg element represents mainly gas transfer resistance and is analogous to a resistor at low frequencies. However, as the cathodic air feed rate was ($\lambda = 2.5$) for the cell studied with H_2/Air and as O_2 was fed at the rate of ($\lambda = 9$) in the case of H_2/O_2 feed, and H_2 feed in both cases was at the rate of ($\lambda = 1.35$), the gas transfer resistance or the corresponding Warburg impedance was not taken into account while determining the values for the equivalent circuit. With this assumption, a modified Randles circuit (shown in Fig 3.6) was used to interpret the EIS spectra obtained at different load currents. The low frequency x-axis intercept was taken as R_{ct} or R_p (polarisation resistance).

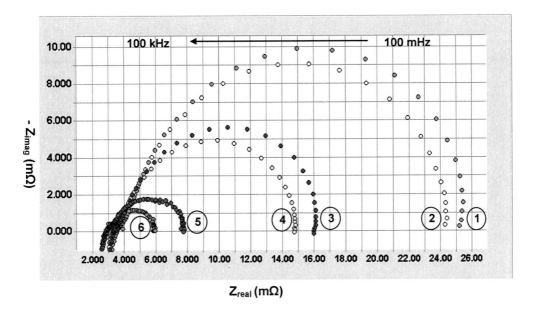

Figure 3.19: EIS spectra of the HT PEMFC Single Cell when operated with H_2/O_2 and H_2/Air (Curve 1 (light blue) indicates H_2/Air operation at 20 mA/cm² of load, ii) Curve 2 (light yellow) represents H_2/O_2 operation. iii) Curve 3 (dark brown) represents 200 mA/cm² and H_2/Air operation and iv) Curve 4 (light green) H_2/O_2 operation; v) Curve 5 (dark pink) represents 400 mA/cm² and H_2/Air operation whereas Curve 6 (bright blue-green) represents H_2/O_2 operation at 400 mA/cm².

In Fig 3.19, bigger semi circle (curve 1) with light blue dots indicates H_2/Air operation at 20 mA/cm² of load, ii) curve 2 with light yellow dots represents H_2/O_2 operation at the same load current. iii) curve 3 with dark brown dots represents 200 mA/cm² and H_2/Air operation and iv) curve 4 with light green dots represents H_2/O_2 operation on the same load, v) curve 5 with bright magenta dots represents 400 mA/cm² and H_2/Air operation whereas curve 6 with bright blue-green dots represents H_2/O_2 operation at 400 mA/cm². As can be seen from Fig 3.19, as the load current increases, the difference in R_p also increases. The values of R_p are higher for air operation and are respectively lower for oxygen operation. As for the membrane resistance or the high frequency intercept, there was no big difference observed from the nyquist plot of the impedance spectrum shown in Fig 3.19. Table 3.7 shows the membrane resistance values at low and high current regions which are more relevant, taking the 0 - 400 mA/cm² as the practical operating regime of a HT PEMFC considered in this work.

Table 3.7: Cell resistances with H_2/Air and H_2/O_2 cell operation (170°C)

H₂/Air operation		H₂/O₂ operation	
Load current (mA/cm²)	Cell resistance (mΩ)	Load current (mA/cm²)	Cell resistance (mΩ)
40	3.46	40	3.46
400	2.91	400	2.83

As can be seen from Table 3.7, the Cell resistance was 3.46 mΩ at a load current of 40 mA/cm², whereas the same fell to 2.91 mΩ at a load current of 400 mA/cm² when the cell was operated with H_2/Air. When operated with H_2/O_2, the same at 40 mA/cm² was 3.46 mΩ and the resistance dropped to 2.83 mΩ at 400 mA/cm². That is to say that on 40 mA/cm² of load, there was no change in electrolyte resistance when the oxidant was changed from air to oxygen. Whereas, at 400 mA/cm² of load, there was a difference of 0.08 mΩ in membrane resistance between H_2/O_2 and H_2/Air operations. By taking this 0.08 mΩ as tolerance in measurements,   one can conclude that there was no change in electrolyte resistance when the oxidant is changed from pure oxygen to air. But the fall in membrane resistance from 3.46 mΩ to 2.91 mΩ (in case of H_2/Air operation) or from 3.46 mΩ to 2.83 mΩ in case of H_2/O_2 operation, as the load current was raised from 40 mA/cm² to 400 mA/cm², could be mainly attributed to the improvement in proton conducting ability of the electrolyte under conditions of cell hydration, aided by the water produced in the fuel cell reaction (the cell temperature being maintained constant at 170°C).

From the EIS spectra obtained from the HT PEMFC single cell operated at 170°C with H_2 and Air with feed rates of (λ = 1.35) and (λ = 2.5) respectively, the high frequency cell resistance (which is mostly dominated by membrane resistance) is represented as R_{el}, the charge transfer resistance R_{ct} which were determined on the basis of the modified Randles model are shown in Table 3.8. The frequencies (f_1) and (f_2) stand for high and low frequencies respectively, at which R_{el} and R_{ct} were recorded respectively, by the EIS analyser.

Table 3.8: Cell resistances with (highly doped membrane) (170°C)

Load current	Current density	R_{el}	Frequency (f_1)	R_{ct}	Frequency (f_2)
(Amps)	(mA/cm²)	(mΩ)	(Hz)	(mΩ)	(Hz)
1	20	3.82	1250	43.54	0.100
2	40	3.46	1250	26.56	0.100
4	80	3.32	1250	16.60	0.125
6	120	3.19	1250	12.94	0.125
8	160	3.10	1250	11.00	0.125
10	200	3.05	1250	9.76	0.159
12	240	3.00	1000	9.22	0.199
14	280	2.97	1000	8.57	0.159
16	320	2.97	1004	8.27	0.100
18	360	2.94	1127	7.91	0.100
20	400	2.91	1000	7.84	0.125

Furthermore, the frequencies (f_1) at which R_{el} was determined varied between 1 kHz and 1.25 kHz as can be seen from Table 3.8. Also, the frequencies (f_2) at which R_{ct} was determined varied between 100 mHz and 125 mHz. While there was no appreciable change in cell resistance, while going from 20 mA/cm² to 400 mA/cm² of load current, the charge transfer resistance R_{ct} fell from 43.5 mΩ to 7.8 mΩ, which is not surprising.

Secondly, tests were performed to study a lowly doped membrane containing HT PEMFC (with a TPS/PA based MEA) single cell at 180°C with H_2/O_2 at (λ = 1.35 and λ = 9) respectively. These lowly doped membranes have their ionic conductivities close to 0.08 S/cm [88] at 190°C, which is about 3.0 times lower than that of highly doped BASF (Celtec P®) membranes. Also, operating them at higher temperatures is desired as proton conductivity is enhanced further. To minimize the effect of gas diffusion resistance (thereby minimizing the Warburg impedance), this cell was operated with H_2/O_2. However, the primary intention was to study the development of R_{el} at increasing load currents.

Following the same procedure explained earlier, employing the very same cell assembly, after connecting the HT PEMFC single cell to the EIS analyser from Gamry® coupled to the electronic load from Basytec®, at load currents of 0 – 400 mA/cm², applying the scanning signal of 0.1 A (a.c), EIS spectra were obtained in the Galvanostatic mode, in the 100 mHz to 100 kHz frequency range. Figure 3.20 depicts nyquist plots pertaining to high frequency impedance of a HT PEMFC single cell containing a lowly doped membrane. The results are summarized in Table 3.9.

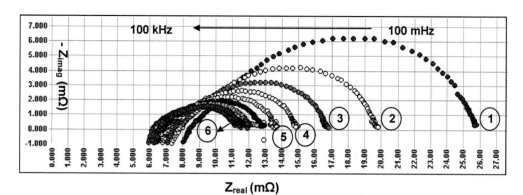

Figure 3.20: Nyquist plot of HT PEMFC single cell (with a lowly doped membrane) impedance at 180°C, when operated with H_2/O_2: Curve 1 indicates operation at 80 mA/cm² of load, ii) Curve 2 represents 120 mA/cm², iii) Curve 3 represents 160 mA/cm², iv) Curve 4 represents 200 mA/cm², v) Curve 5 represents 240 mA/cm² and vi) Curve 6 (dark Magenta) represents 400 mA/cm² load operation. vi) Curves between 5 and 6 represent 280 mA/cm², 320 mA/cm² and 360 mA/cm² in their respective order.

Table 3.9: Cell resistances (lowly doped membrane) (180°C)

Load current	Current density	R_{el}	Frequency (f_1)	R_{ct}	Frequency (f_2)
(Amps)	(mA/cm²)	(mΩ)	(Hz)	(mΩ)	(Hz)
0	0.001	17.21	2239	212.77	0.1
1	20	11.16	1995	67.08	0.1
2	40	9.70	1995	33.89	0.1
4	80	8.39	1779	17.28	0.1
6	120	7.73	1995	12.09	0.1
8	160	7.44	1779	9.39	0.1
10	200	7.16	1995	7.72	0.1
12	240	6.88	1995	6.83	0.1
14	280	6.83	1995	6.00	0.1
16	320	6.66	1995	5.82	0.1
18	360	6.53	1779	5.42	0.1
20	400	6.23	1995	5.18	0.1

Applying the modified Randles circuit model to the impedance spectra obtained, the values of R_{el} and R_{ct} were determined. Curves 1 to 6 in Fig 3.20 indicate Single cell operation at 80 mA/cm² to 400 mA/cm² with regular intervals of 40 mA/cm². As the curves are quite large for load currents lower than 80 mA/cm², they are not shown in Fig 3.20. However, the important values are shown in Table 3.9. Further, the frequencies (f_1) at which R_{el} was determined varied between 1.78 kHz and 2 kHz, except the case of very low current close to 0, where (f_1) is 2.2 kHz. Interestingly, all the (f_2) values at which R_{ct} was determined were 100 mHz.

It can be seen from Table 3.9, that the electrolyte impedance fell from 17.2 mΩ at 0 mA/cm² down to 11.1 mΩ at 20 mA/cm² and then to 6.2 mΩ at 400 mA/cm². Furthermore, it can be observed that the improvement in membrane conductivity is much more pronounced at higher loads in these lowly doped systems, compared to the previously discussed highly doped systems. In other words, water **does** play a major role in these lowly doped systems when compared to highly doped ones, (resulting in lowering of membrane resistance at higher load currents), although the absolute value of membrane conductivity is about 2.5 times lower compared to the highly doped systems. Also, the water uptake of membrane material is usually high (which is the case with PBI as well as non-PBI materials). For instance, Li et al [58] have enumerated that the water uptake of PBI could be as high as 15%. Whereas, the water uptake of the electrolyte (PA) may not be dominant in an operating window of 0 to 400 mA/cm². That is to say that in the case of a highly doped membrane where most part of the electrolyte is PA rather than PBI material, water **does not** influence membrane' conductivity, or they behave as anhydrous proton conductors. Whereas in the case of lowly doped systems, where the polymer content is more than that of the doping agent PA, water **does** influence membrane conductivity, decreasing membrane resistance with increasing loads.

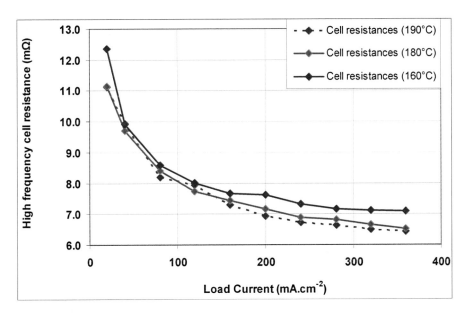

Figure 3.21: Cell resistance at different loads in a lowly doped system: (i) Red curve: operation at 160°C, (ii) Green curve: operation at 180°C and (iii) Dark red curve: operation at 190°C.

In a similar manner, the single cell containing lowly doped membrane was operated at 160°C and 190°C, with H_2/O_2 feeds at (λ = 1.35 and λ = 9) respectively and the EIS spectra thus obtained were fit into modified Randles model as in the previous case. The development of cell resistance and the charge transfer resistance in the case of cell operation at 160°C, 180°C and 190°C is depicted in Fig 3.21 and Fig 3.22. As discussed earlier, the resistance contribution from all the cell components except the membrane (se Table 3.4) for the single cell discussed here was measured to be 0.875 mΩ. Except for this minor contribution from the rest of the cell components, the high frequency cell resistance consists of membrane resistance as is expected. Furthermore, as can be noticed from Fig 3.21, the fall in membrane resistance follows the same trend when operated at 160°C, 180°C and 190°C. Up to a load current of about 150 mA/cm², there is a dramatic fall in membrane resistance, after which, the fall in membrane resistance is not significant. Water production and the subsequent hydration of the lowly doped membrane is the dominant mechanism.

Fig 3.22 shows the development of charge transfer resistance in a HT PEMFC single cell equipped with a lowly doped membrane containing MEA. At the operated temperature regions, the fall in R_{ct} took a similar trend in all the cases (160°C, 180°C and 190°C). The fall in R_{ct} is expected as the load current increased.

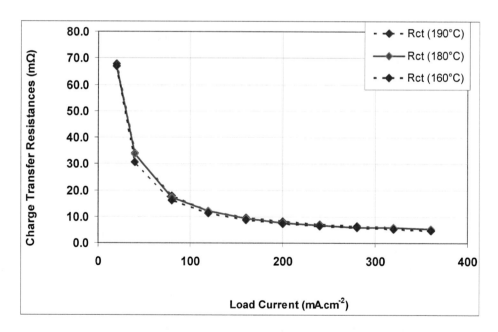

Figure 3.22: R_{ct} from single cell HT PEMFC equipped with a lowly doped MEA: (i) Red curve: operation at 160°C, (ii) Green curve: operation at 180°C and (iii) Blue curve: operation at 190°C.

Furthermore, separate tests were performed with (i) a highly doped and (ii) a lowly doped MEA in proton pump mode and fuel cell mode by the author [89] and found that the influence of water on membrane conductivity is higher in case of lowly doped membranes. The details are beyond the scope of this work. In conclusion, the development of nyquist plots in case of highly doped as well as lowly doped membranes may be schematically depicted as shown in Fig 3.23 and Fig 3.24 respectively, highlighting their spectral signatures.

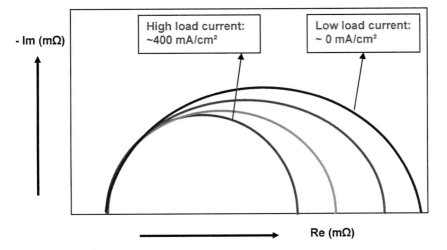

Figure 3.23: Nyquist plot (representative) of cell impedance in a highly doped system: Blue curve – low current; red curve – higher current.

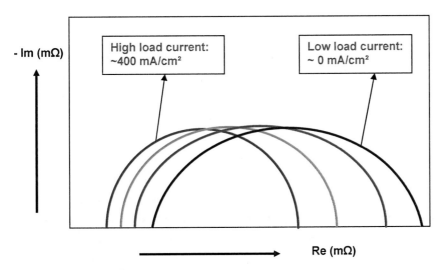

Figure 3.24: Nyquist plot (representative) of cell impedance in a lowly doped system: Blue curve – low current; red curve – higher current.

3.1.3 Fuel transport overvoltage

The loss in cell performance caused due to fuel transport limitations at higher loads was discussed in section 2.1.8. It was also shown that Eq. 30 could be used to calculate these losses at different loads, after finding out the value of the limiting current (i_L). This simple approach was used to ascertain the fuel transport overvoltage in the single cell HT PEMFC containing a highly doped Celtec P 2000 MEA studied here.

Oxygen diffusion into reaction sites is a known problem pertinent to phosphoric acid fuel cells. This is because, the electrolyte in these cells is a liquid which virtually covers most part of the cell including the porous electrodes. The oxygen reduction reaction (ORR) of these cells is therefore limited by slow oxygen diffusion and slow dissolution into the electrolyte on its way to a flooded or partially flooded electrode. As the electrolyte used in HT PEMFCs is basically similar to the ones used in PAFCs, HT PEMFCs are not immune to this problem. As a comparison, LT PEMFCs (conventional low temperature ones) do suffer from slow oxygen diffusion into reaction sites, owing to the presence of liquid water some times flooding the porous electrode. Therefore, presence of water (which is necessary for adequate proton conduction) in LT PEMFCs could offer similar problems as in HT PEMFCs where liquid or gel based electrolyte is used. Furthermore, when air is used as an oxidant instead of pure oxygen, the ORR in PEMFCs is limited by the dissociation of oxygen from the bulk flow (of air) before it reacts with protons at the TPB and this phenomenon is relevant to both low and high temperature PEMFCs. Much effort was dedicated to understanding these performance limiting processes by Scherer' group [90].

Recently, Kamat et al [91] have experimentally investigated the slow oxygen diffusion into the reaction sites of a HT PEMFC at Volkswagen, Germany. They used Chronoamperometry method by studying transient response in cell voltage with a step change in current, especially from very low to very high, and had performed Cottrell experiments to validate their method. They concluded that, diffusion coefficient and solubility of oxygen is about 1000 times slower in acid electrolyte, when compared to a dry cell (without any liquid electrolyte).

In the current work, after feeding H_2 and air at stoichiometries of 1.35 and 2.5 respectively into anode and cathode of the HT PEMFC single cell studied here, load current was drawn until the cell potential had dropped to zero volts and this load current value was considered as the limiting current. This i_L value for the HT PEMFC single cell studied here at 180°C was determined to be 1600 mA/cm². The same at 130°C was 1480 mA/cm². The fuel transport overvoltages which were calculated according to Eq.30 are summarized in Table 3.10.

Table 3.10: Cell voltage loss caused due to mass transport limitations (HT PEM Single Cell)

Load current	Mass transport overpotentials at different temperatures and loads (Volts)					
	130°C	140°C	150°C	160°C	170°C	180°C
mA/cm²	403.15°K	413.15°K	423.15°K	433.15°K	443.15°K	453.15°K
20	0.000473	0.000484	0.000459	0.000469	0.000480	0.000491
100	0.002430	0.002491	0.002353	0.002409	0.002464	0.002520
200	0.005043	0.005169	0.004869	0.004984	0.005099	0.005214
300	0.007869	0.008065	0.007571	0.007750	0.007929	0.008108
400	0.010946	0.011217	0.010490	0.010737	0.010985	0.011233

However, the limiting current (i_L) depends on fuel/oxidant stoichiometries as well as on their partial pressures. As the fuel/oxidant feeds at ($\lambda = 1.35$ and $\lambda = 2.5$) respectively are taken as a standard in this work, also as the observed pressure drops were 2 mbar and 4 mbar respectively on the anode and cathode sides of the single cell, the determined (i_L) values are very specific to the single cell studied here. At fuel/oxidant feeds below the λ values suggested here, the Warburg impedance or gas transfer resistance must be included in the mass transport overvoltages, effectively using the Randles model (See 3.1.1).

3.1.4 Single Cell operation at different temperatures

This section details the performance analysis of a HT PEMFC single cell at different temperatures and load currents. A HT PEMFC was constructed with cell components as shown in figures 3.1 and Fig 3.11 and was operated with H_2/Air as reactant gases, at different load currents. The cell was heated using heating elements inserted into the stainless steel endplates, which were controlled by Lab View® program. Details concerning the test stand were explained in chapter 3.1 (Fig 3.2). It took about 40 minutes to heat the single cell to 160°C from room temperature (20°C). Fig 3.25 depicts the single cell used to perform these tests. MEA used was BASF's Celtec P 2000, with an active area of 50 cm². The heat insulating cover that was wrapped around the single cell was removed to take the picture shown.

Figure 3.25: Single cell (HT PEMFC) characterization in a test stand

With feed gas stoichiometries of 1.35 and 2.5 respectively for H_2 and air at 1 atm, load current was drawn from the said single cell at different operating temperatures. The pressure drop across the anode side flow field was 2 mbar whereas on the cathode side it was 4 mbar. The performance of the single cell from 20 mA/cm² (or 1 amp) to 400 mA/cm² (or 20 amps) at different cell operating temperatures is depicted in Fig 3.26.

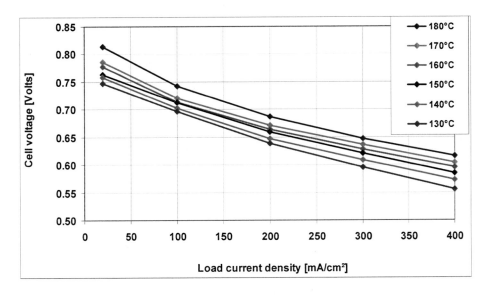

Figure 3.26: Performance of a single cell (HT PEMFC) from 1 Amp - 20 Amp. (i) Brown curve: cell operation at 130°C, (ii) Red curve: operation at 140°C, (iii) Black curve: operation at 150°C, (iv) Green curve: cell operation at 160°C, (v) Magenta curve: operation at 170°C and (vi) Blue curve: cell operation at 180°C.

The OCV of the single cell at different temperatures is not shown in the above figures. However, they are discussed in section 3.1.5. It can be seen from the above figure 3.26 that the cell voltage falls gradually with a gradual rise in load current and also with a gradual fall in cell temperature, as is expected. As was discussed in the earlier sections (Fig 3.16), conductivity of the membrane considered here is better at higher temperatures. As the operation of the cell up to 400 mA/cm² implies the cell operation in the ohmic region of a fuel cell as illustrated in Fig 2.2, the fall in cell performance could be mainly attributed to a fall in membrane conductivity as the cell temperature falls from 180°C to 130°C. Furthermore, the kinetics of ORR at the cathode of a PEMFC are normally enhanced as the temperature increases, resulting in higher cell performance. At higher currents (> 800 mA/cm²), mass diffusion impedances dominate. The mass transport overvoltages were discussed in the previous section, whereas the kinetic overvoltages are discussed in the following section. The single cell performance is summarized in the form of a polynomial relationship between cell voltage and cell's current density for various operating temperatures of HT PEMFC and is shown in Eq 38.

<table>
<tr><td>$V = ax^2 - bx + c$</td><td>where</td><td rowspan="4">Eq. 38</td></tr>
</table>

$V = ax^2 - bx + c$ where

V = Cell Voltage (Volts)

x = current density (mA/cm²)

a,b,c = constants (shown in Table 3.11)

Eq. 38

Table 3.11: Coefficients of polynomial equation representing HT PEMFC performance

Cell Temp (°K)	Cell Temp (°C)	a	b	c
453.15	180	0.0000009187	0.0008917526	0.8281591572
443.15	170	0.0000007681	0.0007867054	0.7979803081
433.15	160	0.0000007476	0.0007790040	0.7897600902
423.15	150	0.0000004979	0.0006753254	0.7758110513
413.15	140	0.0000005974	0.0007333852	0.7714410056
403.15	130	0.0000004563	0.0006956410	0.7611188654

It is possible to reconstruct performance of the HT PEMFC discussed in this work from Eq 38 and Table 3.11 with a negligible error. Operating the HT PEMFC at temperatures close to 180°C might especially be attractive when the fuel considered is a reformate (and not pure H_2), as the enhanced kinetics offer higher tolerance to impurities such as CO, H_2S and the like. Operating the cell at lower temperatures (~ 130°C) might be interesting during the start up period, as these HT PEMFCs need to be first heated up from room temperature to the required cell temperature. Operating the cell at 130°C is not attractive from the point of view of cell performance and tolerance to fuel impurities. Whereas operating the cell close to 180°C may not be the best from the point of view of electrolyte retention. Thus, an operating cell temperature of 160°C might be a viable compromise between cell performance and tolerance to impurities, but offering higher durability. Some of these aspects are discussed in the following sections with more details. Therefore, taking the HT PEMFC operation at 160°C as a reference, the rise in cell performance (denoted by a rise in cell voltage) per 1° rise in cell temperature when the load current is raised from 5 amp (or 100 mA/cm²) to 20 amps (400 mA/cm²) in four steps is depicted in Fig 3.27. Similarly, the fall in cell voltage per 1° fall in its temperature also is depicted in Fig 3.27.

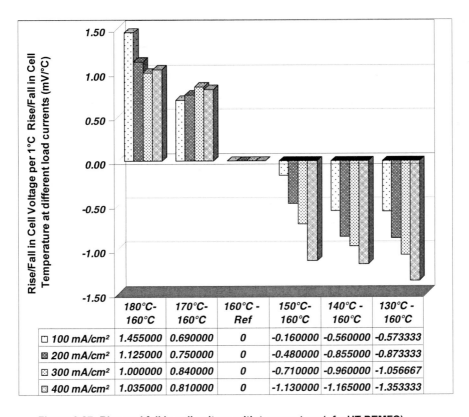

	180°C-160°C	170°C-160°C	160°C - Ref	150°C-160°C	140°C - 160°C	130°C - 160°C
☐ 100 mA/cm²	1.455000	0.690000	0	-0.160000	-0.560000	-0.573333
▨ 200 mA/cm²	1.125000	0.750000	0	-0.480000	-0.855000	-0.873333
▩ 300 mA/cm²	1.000000	0.840000	0	-0.710000	-0.960000	-1.056667
▢ 400 mA/cm²	1.035000	0.810000	0	-1.130000	-1.165000	-1.353333

Figure 3.27: Rise and fall in cell voltage with temperature (of a HT PEMFC)

It can be noticed from Fig 3.27, that the fall in cell performance is commensurate with a rise in load current (from 100 mA/cm² to 400 mA/cm²). Also, a very high loss in cell performance can be observed at 400 mA/cm² when the cell temperature is lowered from 160°C. For instance, it was shown that, going from 160°C to 170°C (ΔT= 10 K), at a load current of 100 mA/cm², the performance *gain rate* was 0.69 mV/°C or a total performance *gain* of 6.9 mV (0.69 * 10). Whereas, at 400 mA/cm², under same conditions (160°C to 170°C), the performance gain rate was 0.81 mV/°C or a total gain of 8.1 mV. Also, while going from 160°C to 180°C (ΔT= 20 K), the performance gain rate was 1.45 mV/°C (100 mA/cm²) and it was 1.03 mV/°C (400 mA/cm²), which translates to a total cell performance gain of 29 mV (1.45 * 20) in the first case and a gain of 20.6 mV (1.03 * 20) in the second case. Similarly, while going from 160°C down to 130°C (ΔT= -30 K), it was observed that at 100 mA/cm², the performance loss rate was 0.57 mV/°C, and at 400 mA/cm², it was 1.35 mV/°C. That implies a total performance loss of 17.19 mV (0.57 * 30) at 100 mA/cm² and a total loss of 40.59 mV (1.35 *

30) at 400 mA/cm². As a comparison, in a typical PAFC, operating with H_2 and Air (and operating at around 180°C), at a load current of 250 mA/cm², the reported gain in cell performance (with a rise in cell temperature) was about 1.15 mV/°C, as reported by a Department of Energy (DoE) study [92].

3.1.5 Open circuit voltage (OCV) of a HT PEMFC

Open circuit voltage (OCV) of a HT PEMFC is one of the important points to discuss. In chapter 2.1 of this work, reasons for lower theoretical cell voltages in PEMFCs operating at temperatures above boiling point of water were explained. Table 13 in (Appendix I) shows the ideal theoretical OCV of a PEMFC operating at different temperatures (with H_2/O_2). When air is used in stead of pure oxygen in the fuel cell reaction considered here, a loss of 15.2 mV at 180°C and a loss of 13.5 mV at 130°C can be expected as a result of reduced O_2 partial pressure. Furthermore, in the current work, the measured OCV (no load voltage) of the cell was taken into account while deducing activation overvoltages. The difference between the theoretical thermodynamic potential discussed above and the OCV measured at the cell terminals can be attributed to (a) loss caused due to lower O_2 partial pressure, (b) loss caused due to product water partial pressure, based on the Nernst relations shown in Eq. 21(a-c), (c) loss caused due to leakage currents of the membrane considered at their respective operating temperatures as enumerated by Gasteiger and Weber [44,93] and d) loss caused due to phosphate anionic adsorption on the catalyst sites.

The absolute pressure of the reactant gases fed into the HT PEMFC single cell studied in this work were 1.002 atm and 1.004 atm for H_2 and air respectively. In theory, according to the Nernst equation for cell potential shown in Eq. 21, this would mean a rise in cell voltage of 17.1 µV (130°C) to 19.2 µV (180°C) due to the slight increase in H_2 partial pressure and a rise of 68 – 76 µV (130°C – 180°C) due to the 4 mbar rise in cathode side air partial pressure. Wang et al [94] in 1995, have studied the acid doped PBI cell with H_2 - O_2 and H_2 - Air at 150°C in their laboratory and had reported a decrease in OCV of 45 mV, as a direct result of reduced O_2 partial pressure. This was based on a MEA containing home-made cathode side electrode with a Pt loading of 2 mg.cm^{-2}.

Liu et al [95] have reported higher hydrogen crossover from their HT PEM cell test after 600 hours of operation causing lower OCV as a result. However, at higher stoichiometries used for the single cell studied in this work, (e.g., λ=1.35 for H_2 and λ=2.5 for air), the gas crossover or leakage currents of the membrane are assumed to be negligible. For instance, Plug-power® [96] in accordance with PEMEAs (later known as BASF fuel cells) has indicated leakage currents, in similar membranes, to be in the range of 2-5 mA/cm^2 each for H_2 and O_2. For a total leakage current of 200 - 500 mA (for a 50 cm^2 membrane with 1.88 mΩ high frequency impedance), this would be a loss of 0.377 mV to 0.94 mV. Shamardina et al [97] have reported the loss in cell voltage due to leakage currents to be close to zero, except for current densities higher than 1200 mA/cm^2, based on their in-house MEAs.

As mentioned in section 2.2.1, Qinggang He et al have studied the effect of phosphate anionic adsorption on the electrocatalytic activity of ORR on Pt single crystal surface. At higher concentrations of PA and depending on Pt crystal surface structure, they concluded that PA adsorption on Pt varied. This phosphate anionic adsorption is one reason for lower OCVs observed in PAFCs and HT PEMFCs. Wainright et al have reported the same cause for lower OCVs in cells containing PA doped membranes [98]. As operating HTPEMFCs on OCV for prolonged periods of time is detrimental to its electrochemically active surface area (ECSA), the cell studied here was not operated on OCV for more than 60 seconds. However, the observed OCV in the 90°C – 170°C temperature range was 0.955 V - 0.985 Volts, as depicted in Fig. 3.28. These values are appropriate for relatively new MEAs (at beginning of life or BOL), after which the observed OCV was mostly around 0.8 V, possibly due to phosphate anionic adsorption. Overall, it may be said that in these cells, the thermodynamics of high temperature operation cause the cell's potential to be lower than their LTPEMFC counterparts, whereas the enhanced kinetics facilitated at these very same conditions, almost nullify that effect. And the phosphate anionic adsorption on catalyst sites leads to lower OCVs close to 0.8 V. Furthermore, these cells have 30% more noble metal catalyst at today's state of development, than the low temperature counterparts.

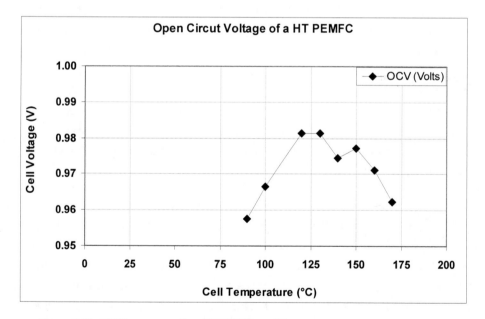

Figure 3.28: OCV in an operating HT PEMFC at different temperatures (H$_2$/Air)

3.1.6 Activation overvoltage in a HT PEMFC

The cell performance loss caused due to slow reaction kinetics at the triple-phase boundary (TPB), which is also called the activation overvoltage, was discussed earlier in section 2.1.6 using the Eq.22. These activation overvoltages for a HT PEMFC single cell operating at different temperatures were determined by deducting the cell voltages corrected for ohmic overvoltages shown in Table 3.5 as well as the mass transport overvoltages shown in Table 3.10 from the cell's OCV. The activation overvoltages (caused due to slow kinetics at anode and cathode side reaction sites) are summarized in Table 3.12. The Table 3.13 shows the corrected cell voltages.

Table 3.12: Activation overvoltages in a HT PEMFC (130°C – 180°C)

Current density	Activation overvoltage (both anodic and cathodic) (Volts)					
mA/cm²	130°C	140°C	150°C	160°C	170°C	180°C
20	0.219105	0.208517	0.203437	0.189278	0.180815	0.153258
100	0.256359	0.250415	0.241426	0.240031	0.233310	0.211327
200	0.297535	0.288542	0.277489	0.273096	0.265950	0.251379
300	0.322599	0.310251	0.300066	0.293569	0.285694	0.274732
400	0.345512	0.328403	0.319027	0.308522	0.301112	0.289353

Table 3.13: Cell voltages corrected for ohmic and transport overvoltages (130°C - 180°C)

Current density	Cell voltages corrected for ohmic and transport overvoltages (Volts)					
mA/cm²	130°C	140°C	150°C	160°C	170°C	180°C
20	0.750895	0.761483	0.766563	0.780722	0.789185	0.816742
100	0.713641	0.719585	0.728574	0.729969	0.736690	0.758673
200	0.672465	0.681458	0.692511	0.696904	0.704050	0.718621
300	0.647401	0.659749	0.669934	0.676431	0.684306	0.695268
400	0.624488	0.641597	0.650973	0.661478	0.668888	0.680647

The HT PEMFC's performance as well as various overvoltages are depicted in Figure 3.29 for cell operation at 180°C. As can be expected, activation overvoltages are the highest losses followed by ohmic losses and fuel transport losses in their respective order. However, at very large current densities (above 1000 mA/cm²), fuel transport overvoltages will be substantial. It can be seen from Table 3.12 and Table 3.13 that the same trend holds good for cell operation at temperatures of 130°C to 180°C, only that the activation overvoltages at lower temperatures are higher. For instance, at 400 mA/cm², the corrected single cell voltage was 0.6806 Volts and the activation overvoltages was determined to be 0.2893 Volts at 180°C. Whereas at the same load current, the corrected cell voltage was 0.6244 Volts at 130°C and the activation overvoltages was 0.3455 Volts. This difference of 56.2 mV can be mainly attributed to slower kinetics when the cell temperature is lowered to 130°C from 180°C, at 400 mA/cm².

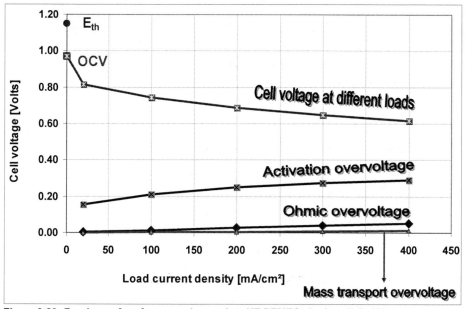

Figure 3.29: Break-up of performance losses in a HT PEMFC single cell (H₂/Air

3.1.7 Determination of Kinetic Parameters

As was discussed in section 2.1.6 (Eq.22), the fuel cell's activation overvoltage is a function of variables such as exchange current density (i_0), the electron transfer co-efficient (α) and operating temperature (T). In the Eq. 23, "b" stands for Tafel slope, which is the slope of the Tafel curve at its respective temperature. Fig 3.31 depicts Tafel slopes of the curves shown in Fig 3.30 (plot of activation overpotential versus logarithm of load current). The pivot points for determination of Tafel slopes shown in Fig 3.31 were always the co-ordinates corresponding to (activation overvoltages at higher current densities of ln(i) > 5 mA/cm² versus higher load currents). Usually, there will be a big change in slope of the Tafel curve towards lower load currents.

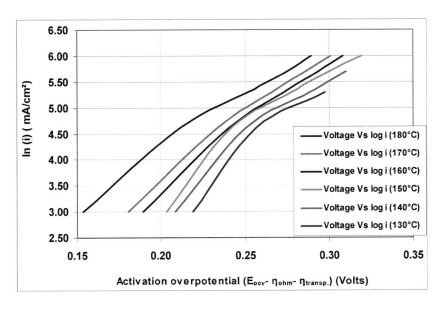

Figure 3.30: Tafel curves of a HT PEMFC single cell (H₂/Air)

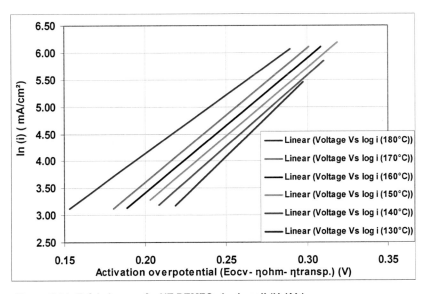

Figure 3.31: Tafel slopes of a HT PEMFC single cell (H₂/Air)

The values of "i_0" were determined from the y-axis intercept of Tafel slopes (Fig 3.31 at zero activation overpotentials (at x = 0). The values of Tafel slopes, and "i_0" at different cell operating temperatures are shown in Table 3.14.

Table 3.14: Tafel slopes and (i_0) of a HT PEMFC (130°C – 180°C)

Temperature (°C)	Exchange current density "i_0" (mA/cm²)	Tafel slope (mV/decade)	Transfer coefficient (α)
180	8.313E-01	106.48	0.422
170	2.535E-01	92.73	0.474
160	2.046E-01	92.34	0.465
150	1.616E-01	91.63	0.458
140	1.037E-01	88.03	0.465
130	4.042E-02	79.02	0.506

From the Tafel slopes shown in Table 3.14 at different cell operating temperatures, the "α" values were calculated from Eq.22 shown in section 2.1.6. This was done by using all the known values of "η_{act}", "i_0", R, T, F and i, in the equation for activation overvoltage given by:

$$\eta_{act} = \left[\frac{RT}{\alpha nF}\right] * \ln\frac{[i]}{[i_0]}$$

For instance, for n=2 (number of electrons transferred in the fuel cell reaction), the value of "α" at 180°C was found to be 0.42 and at 130°C, it was 0.50. It can be clearly from Table 3.14 that the exchange current densities are higher at higher temperatures, and this explains why kinetics are enhanced at higher cell operating temperatures. Whereas it was discussed in the previous sections that the thermodynamics of fuel cell reaction cause the cell voltage to fall at higher temperatures.

3.1.8 Determination of Kinetic Parameter (i_0) from EIS spectra

The exchange current density values can be derived from the nyquist plots of the cell impedance (EIS spectra) obtained at very low currents densities, from the charge transfer resistance values. According to Eq.22 in section 2.1.6, the activation over voltage varies logarith-
mically with the ratio (i/i_0). However at low activation losses (η_{act} < 10 mV) which occurs at
very low currents, the expression for activation overvoltage can be simplified to the form
shown in Eq.39, where the η_{act} value varies linearly with the ratio (i/i_0) [99]. Furthermore, at
such low η_{act} values, the activation loss is dominated by slow ORR kinetics and the charge
transfer resistance from the EIS spectrum of the cell impedance follows the relation shown in
Eq.40. The exchange current densities thus derived are also termed as "Apparent exchange
current densities (AECD)".

In reality, it is not easy to obtain R_{ct} values from the nyquist plots (EIS spectra) of the cell impedance obtained at very low currents, if we were to go by the x-axis (low frequency) intercept of the nyquist plot, which in theory represents R_{ct} when the fuel transport and fuel crossover losses are ultra low. This is due to the Warburg impedance type behaviour of the EIS
spectra at lower loads, which shows infinite diffusion behaviour at lower frequencies and
which does not always have a clear lower frequency x-axis intercept (with zero imaginary impedance). However, from the $(-Z_{imag})_{max}$ values (see section 3.1.1 and Fig 3.4) obtained from
the cell impedance spectra which is assumed to follow modified Randles model, where the
gas resistance is negligible (or the Warburg element is not siginificanly high), the R_{ct} values
can be determined from the relation shown in Eq.41, according to Orazem et al and Barsoukov et al, based on their models described in section 3.1.1.

$$\eta_{act} = \left[\frac{RT}{\alpha nF} \right] * \frac{[i]}{[i_0]}$$	Eq.39
$$R_{ct} = \left[\frac{RT}{\alpha nFi_0} \right]$$	Eq.40
$$-Z_{imag(max)} = \left[\frac{R_{ct}}{2} \right]$$	Eq.41

A single cell HT PEMFC was operated at 160°C – 180°C temperatures fed with H_2 and O_2 of 100 ml/min each, impedance spectra were obtained by operating the EIS instrument, following the very same procedure detailed earlier, at load current values of 50 mA (or 1 mA/cm²). From the $(-Z_{imag})_{max}$ values obtained from these nyquist plots, the R_{ct} values were determined from the relation shown in Eq.41. After substituting these values of R_{ct} in Eq.40, for ($\alpha n = 2$), at different operating temperatures, the exchange current densities were determined. All the values are shown in Table 3.15. In comparison to the exchange i_0 values obtained earlier (Table 3.14) from the Tafel slopes, these i_0 values (Table 3.15) obtained from the EIS spectra at very low loads are larger. For instance, at 160°C, the i_0 values from the EIS spectra are 2.7 times higher, at 170°C they are 7.3 times higher and at 180°C, they are 2.5 times higher compared to the Tafel slope based approach. However, it must be emphasised that these EIS spectra are hard to obtain, depending on the state of the membrane. For instance at temperatures of around 160°C -180°C, operating a HT PEMFC cell with such low loads is not recommended as the degradation of the catalyst would be triggered as detailed in the following section at such high cell voltages (> 0.8 V). Furthermore, PA could get dehydrated (under conditions of high temperature and low load when almost no water is produced in fuel cell) resulting in a fall in ionic conductivity of the electrolyte. This would mean a changing triple-phase boundary which renders it difficult to obtain repeatably such EIS spectra at low currents. Also, it is not recommended to do so, as it would cause further degradation of the catalyst. However, an attempt was made here to understand the impedance spectra at low currents. Tang et al [100] have made similar studies on Celtec P® based MEA, and found the AECD (i_0) values from the corresponding R_{ct} values at OCV. They found that the AECD (i_0) values were 2.64 mA/cm² at 160°C, 5.43 mA/cm² at 200°C and 5.16 mA/cm² at 300°C.

Table 3.15: Exchange current densities (i_0) of a HT PEMFC (160°C – 180°C)

Cell Temp.	R_{ct} (Ω)	I_0 (Amps)	i_0 (mA/cm²)	$(-Z_{imag}$ max)(Ω)
160	0.665198	0.02805483	0.56109656	0.332599
170	0.205898	0.0927297	1.85459398	0.102949
180	0.20438	0.09552649	1.91052977	0.10219

The AECD values shown in Table 3.15 translates to the activation overvoltages of 122 mV at 160°C (400 mA/cm² load), 102 mV at 170°C and 104 mV at 180°C on the same load. These AECD based activation overvoltages are about 2.5 times lower compared to Tafel slope based activation overvoltages discussed above. As the AECD values hold good only for cell operation close to OCV, these differences are expected.

3.2 HT PEMFCs' tolerance to fuel impurities

3.2.1 CO tolerance of HT PEMFCs

When a HT PEMFC is fed with a fuel (such as hydrogen) containing impurities like CO, depending on the current drawn from the cell, some voltage is lost as these impurities present
in the fuel stream could block the reaction sites. Therefore at temperatures of > 160°C, as
the kinetics are improved, adsorption of CO on the catalyst sites of a HT PEMFC will be less
of an issue, rather the dilution of the fuel stream (at CO contents of > 5%) will be an issue.
Whereas, a HT PEMFC operating at temperatures below 160°C, may not offer very high CO
tolerance with a reasonable loss in cell performance. Experiments were carried out to understand CO tolerance of a HT PEMFC (of the type shown in Fig 3.1) containing a Celtec P®
2000 MEA. Fig 3.32 depicts loss in cell voltage of this cell when fed with only 1% CO into the
anode stream. In this case, as discussed in earlier sections, stoichiometries of 1.35/2.5 for
(H_2+CO)/Air respectively were maintained. The reference case is a HT PEMFC being fed
with H_2/Air (without any CO being present in the fuel stream). The upward arrows in Fig 3.32
indicate that the values go beyond the range (on the Y-axis).

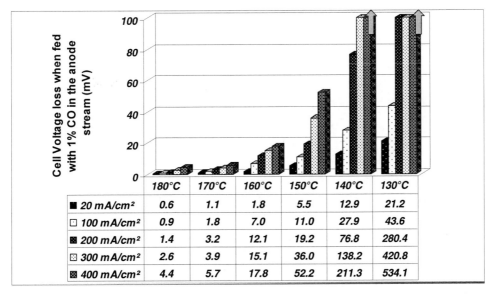

	180°C	170°C	160°C	150°C	140°C	130°C
■ 20 mA/cm²	0.6	1.1	1.8	5.5	12.9	21.2
□ 100 mA/cm²	0.9	1.8	7.0	11.0	27.9	43.6
▦ 200 mA/cm²	1.4	3.2	12.1	19.2	76.8	280.4
▨ 300 mA/cm²	2.6	3.9	15.1	36.0	138.2	420.8
▩ 400 mA/cm²	4.4	5.7	17.8	52.2	211.3	534.1

Figure 3.32: Cell voltage loss when fed with 1% CO into anode stream (H_2+CO/Air: 1.35/2.5)

It can be seen from Fig 3.32 that 1% CO in the fuel stream limits operation of the fuel cell either to higher temperature operation or the cell operation at 140°C to low current density. To

enumerate further, the β values (ratio of overvoltages caused due to CO feed to Tafel slope) were calculated from the relation $(\Delta V)/(RT/nF)$, where ΔV stands for cell voltage loss when fed with 1% CO compared to the case of 100% H_2 feed, R, and F are standard values, n = 2 and T is the cell's operating temperature in Kelvin. These β values were found to vary between 0.031 (at 20 mA/cm²) at 180°C to 0.225 (at 400 mA/cm²) at 180°C. At 160°C, β values were 0.096 (20 mA/cm²) and 0.954 (400 mA/cm²). Whereas at 150°C, β values were 0.302 (20 mA/cm²), 0.603 (100 mA/cm²), 1.053 (200 mA/cm²), 1.975 (300 mA/cm²) and 2.863 (400 mA/cm²). All the β values corresponding to HT PEMFC operation with 1% CO feed into its anode stream at (H_2 + CO) feed rates of (λ = 1.35) and air feed rate of (λ = 2.5) at different operating temperatures and load current densities are shown in Table 3.16. In conclusion, it can be said that operating a HT PEMFC when fed with 1% CO into its anode stream with feed rates of (λ = 1.35) corresponding to β values of less than 1 is recommended to ensure cell operation without significant performance loss.

Table 3.16: Dimension-less β values at different loads and temperatures

Dimension-less β values at different loads and temperatures (for a HT PEMFC operation with 1% CO into its anode) with H_2/Air feeds of λ = 1.35/2.5 respectively				
Temp.	20 mA/cm²	100 mA/cm²	200 mA/cm²	300 mA/cm²	400 mA/cm²
180°C	0.031	0.046	0.072	0.133	0.225
170°C	0.058	0.094	0.168	0.204	0.299
160°C	0.096	0.375	0.648	0.809	0.954
150°C	0.302	0.603	1.053	1.975	2.863
140°C	0.725	1.567	4.315	7.764	11.871
130°C	1.221	2.510	16.143	24.226	30.749

The HT PEMFC single cell's tolerance to CO was further studied at higher concentrations of CO as enumerated further with Fig 3.33. It can be observed that the cell voltage fell to 0.5278 V when the CO feed was 20% from its initial value of 0.6869 V at 180°C, when no CO was present in the fuel gas. More information on CO tolerance of a HT PEMFC at 170°C, 150°C can be found from Figures 1 to 3 in Appendix II at the end of this work. Fig 3.33 depicts cell performance at 180°C at 200 mA/cm² with 20% CO feed in fuel gas.

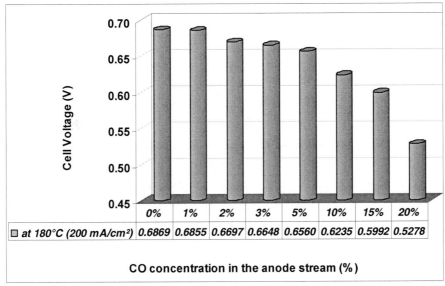

Figure 3.33: CO tolerance of a HT PEMFC at 180°C (H_2+CO/Air: 1.35/2.5)

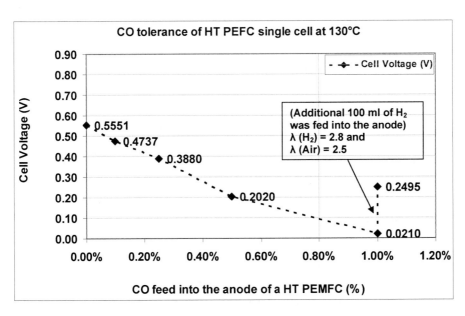

Figure 3.34: CO tolerance of a HT PEMFC at 130°C at 400 mA/cm² (H_2+CO/Air: 1.35/2.5); At 1% CO feed when cell voltage dropped close to zero, 100 ml of H_2 was injected into the anode, keeping the cathodic air feed constant.

CO tolerance of the cell at lower cell temperatures is discussed further. Fig 3.34 depicts the HT PEMFC performance when operated at 130°C with CO feeds of 0 to 1% into its anode stream. As can be observed, with pure hydrogen feed, the cell voltage was 0.55 Volts at 400 mA/cm², whereas with only 0.1% of CO content in the anode stream, the cell voltage fell to 0.4737 V. The cell was fed with pure hydrogen at the same load till it gained back its original cell voltage. When the CO concentration was raised to 0.5%, at the same load, the cell voltage fell to 0.20 V and finally at 1% CO, the cell had lost almost the entire cell voltage. To revive the cell, an additional 100 ml of H_2 (equivalent to a λ = 2.8 on the anode side) was injected into the anode, resulting in the cell voltage of 0.2495 Volts shown in Fig 3.34. Therefore, it may be concluded that a HT PEMFC operating at 130°C, may not offer very high CO tolerance. It is important to bear in mind that most of the reformates (such as methanol, natural gas, propane, etc) contain CO in the % range, when there is no PrOx (preferential oxidation) stage involved. In such a case, starting up of a HT PEMFC at cell temperatures of 130°C might be impractical, where the cell might lose most of its performance. Cell operation at temperatures less than 130°C, might be even more difficult with reformate feeds into a HT PEMFC. Some details of CO tolerance at 130°C may be found from Figure 4 in Appendix II. Performance characteristics of a HT PEMFC when fed with different reformates is being discussed in section 3.3 with more details.

In conclusion, HT PEMFC's tolerance to CO (carbon monoxide) of 1% to 20% in the anode stream was studied. It was observed that at 180°C, a HT PEMFC fit with a Celtec P® 2000 MEA could offer a CO tolerance of up to 20% with a loss of about 150 mV (at 200 mA/cm² of load), whereas the same could not even stand 0.5% CO at 130°C. While designing HT PEMFC-liquid or gaseous fuel reformer coupling, care must be taken to avoid reformate streams at cell temperatures below 130°C (with CO concentration of up to 0.1%). Furthermore, HT PEMFC single cell's performance was studied when fed with 1% CO with its anode stream of hydrogen in the 130°C – 180°C temperature range and in the 20 mA/cm² - 400 mA/cm² load range, as it is more interesting to the reformer-coupled-HT PEMFC system configurations. It was shown that at temperatures of more than 170°C, 1% CO in the anode stream did not result in a significant cell voltage loss, whereas at temperatures below 150°C, the voltage loss was considerable. For instance, at 180°C, with 1% CO feed into the anode stream of hydrogen (λ = 1.35), cathode being fed with air (λ = 2.5), while going from 20 mA/cm² to 400 mA/cm² load, the cell had lost 0.6 mV to 4.4 mV respectively (compared to the case of pure hydrogen feed at the same (λ of 1.35), which is negligible. But at 150°C, with 1% CO feed into the anode stream of hydrogen (λ = 1.35) and under the very same conditions (as at 180°C), while going from 20 mA/cm² to 400 mA/cm², the cell voltage loss was 5.5 mV to 52.2 mV respectively, compared to the case of pure hydrogen feed, which is considerable. Furthermore, at 130°C, with 1% CO feed into the anode stream, under the very same conditions as in the previous two cases, while going from 20 mA/cm² to 400 mA/cm² load, the cell had lost 21.1 mV to 534.1 mV respectively, compared to the case of pure hydrogen feed which is significant.

3.2.2 HT PEMFC's tolerance to Ammonia (NH₃), Hydrogen Sulfide (H₂S) and Methanol (CH₃OH)

It is well known that many impurities other than CO could be present in reformates that are being fed into a HT PEMFC, depending on the quality of fuel considered. For instance, most of LPG bottles might have some Hydrogen Sulfide (H_2S) present in it, although in a ppm range. Similarly, ammonia might be present in a natural gas based reformate and some amount of unconverted methanol might be present in a methanol reformate. When a HT PEMFC is to be coupled to any reformer one has to take the various reformate constituents into account. Sometimes, there might be no H_2S present in the reformate, as a desulphurisation step employed in a reformer might completely eliminate the same. However, for all practical purposes, it is of interest to understand the HT PEMFC's tolerance to impurities.

With regard to the impurities present in the atmospheric air, it may be noted that impurities such as NOx, SO_2, HC (hydrocarbons), CO_2, ground level O_3, etc might affect fuel cell performance [101]. It is beyond the scope of this work to go into further details of fuel cell performance degradation with impurities present in the oxidant. Much can be learned from PAFCs, as they are an established technology. It was observed that about < 50 ppm of hydrogen sulfide and carbonyl sulphide (H_2S + COS) present in a reformate might be tolerated in a PAFC operating at 190°C – 210°C and at 9.2 atm at 325 mA/cm², without a destructive loss in cell performance [102]. The same report also concluded that a tolerance of < 20 ppm of H_2S could be expected from a PAFC. But it is important to note that the combined effect of H_2S+CO could cause a pronounced performance loss in a PAFC as reported by James Douglas [103]. In their work they found that only more than 240 ppm of H_2S had triggered a very high cell performance loss (i.e., when no CO was fed into a PAFC operating at 190°C and a load current of 200 mA/cm²). Whereas, with a CO feed of 10% into the anode stream, about 160 ppm of H_2S had resulted in a substantial fall in cell performance.

With regard to HT PEMFCs operating with highly doped MEAs, such as Celtec P® MEAs from BASF, Schmidt et al had indicated that about 10 ppm of H_2S at 180°C, even with the presence of CO in the reformate gas did not result in any irreversible cell degradation. In their 3500 hour test with a single cell containing Celtec P 1000 MEA, fed with (60% H_2, 2% CO, 21% H_2O, 17% CO_2 and 5 ppm H_2S), operated at 180°C, with a load current of 200 mA/cm², they had reported a performance loss of about 100 mV during the entire 3500 hours of cell operation [104].

NH_3 tolerance of HT PEMFCs

From the experiences made from PAFCs, Minh and Lundberg studies [105,106] show that NH_3 present in the fuel stream (from a reformer) reacts with H_3PO_4 to form a phosphate salt known as $(NH_4)H_2PO_4$. A concentration of less than 0.2 mole% of the said phosphate salt must be maintained to avoid unacceptable performance loss. Issues related to the presence of ammonia in the fuel stream are electrolyte related rather than catalyst related. Formation of NH_3 in natural gas reformers in the event of a nitrogen purge can not be ruled out. Gerhard Ertl [107], in his Nobel Prize winning scientific work, elucidated the dynamics of NH_3 formation from N_2 and H_2 on catalyst surfaces [Figure 5 in Appendix II]. Any trace amounts of ammonia present in the reformate stream, will have the potential to react with the PA based electrolyte employed in HT PEMFCs. However, Schmidt [108], has indicated that NH_3 up to several percent could be tolerated by HT PEMFCs containing Celtec P ® based MEAs from BASF.

CH_3OH tolerance of HT PEMFCs

Performance of a HT PEMFC in the presence of methanol related species might be of interest for potential users of HT PEMFC stacks. Investigations have been performed with 0.15% of methanol mixed with H_2 and air, employing the same MEA (Celtec P® 2000 from BASF) used in the other tests and found that there was virtually no influence on cell performance when the cell was operated at 160°C – 180°C from 20 mA/cm² to 400 mA/cm². Cell performance with pure (99.999%) H_2 and air being the reference case. It is believed [109] that about 2% concentration of CH_3OH should not be exceeded, to maintain acceptable cell performance. However, many attempts were made to use methanol as fuel in HT PEMFCs, avoiding fuel reforming step. For instance, BASF, Germany and PSI, Switzerland have worked on Celtec V (which was based on PBI/PVPA or PBI/Polyvinylphosphonic acid composite membrane) for use in direct methanol (vapour) fuel cell.

Gubler [110] et al have found that Celtec V has about 50% lower methanol crossover com-pared to those of Nafion 117 based membranes. However, the resistance of Celtec V based membrane was found to be 30% higher than a Nafion 117 based one. Furthermore, they elu-cidated that the cathode performance was notably inferior above 400 mA/cm² when Celtec V was used in a DMFC. This phenomenon was not clearly understood and it was proposed that the patterns of water distribution and a possible poisoning of the cathode catalyst by the con-stituents of PBI/PVPA based membrane could be the reason for that inferior performance. Furthermore, a durability test of about 500 hours with Celtec V had shown that there was an increase of 18% in membrane resistance and some possible MEA deterioration. Calundann and Henschel [111] have compared the performance of Celtec V and Nafion 117 based MEAs when fed directly with methanol and water mixtures as in a DMFC.

3.3 HT PEMFC stack under the influence of reformates

Performance of a HT PEMFC stack when operated with three types of synthetic reformates also is discussed, in the following sections, after providing some insights into basic fuel reforming approaches.

3.3.1 General motivation for coupling HT PEMFCs to reformers

Manufacturers of CHP (combined heat and power) units, UPS (uninterrupted power supply) units, the automobile sector, and various mobile electronics sectors are definitely eager to go in for this technology owing to many advantages this HT PEMFC has to offer when compared to the classical LT PEMFC technology. For instance, a bulky PROX (preferential oxidation) reactor for CO clean up could be eliminated (in the case of HT PEMFCs), which is a must when LT PEMFC were considered. This reduces size and cost of the entire system (consisting a reformer and a fuel cell stack). Secondly, water management (humidification of reactant gases) is not an issue; thereby size and cost of humidifiers and their controls can be eliminated. Thirdly, there can be a synergetic heat management in a reformer coupled HT PEMFC stack. For instance, methanol or ethanol based fuels could be reformed at temperatures close to 200°C and a typical HT PEMFC's operating regime is 160°C to 180°C, therefore hot reformate gases exiting from a methanol reformer could be fed directly into a HT PEMFC stack. On the other hand, heat produced in the stack could sustain the entire system at the required temperature. This implies higher total system efficiency for HT PEMFCs. Nevertheless, there exist some challenges to overcome, such as start up of a HT PEMFC system connected to reformers, thermal and load cycling with reformate feeds, etc. Volkswagen [112] had demonstrated its 45 kW electric motor driven small FCV (called Space up! Blue), in which the lithium ion batteries used could be charged by a HT PEMFC stack. UltraCell®, USA has developed a 25 Wel HT PEMFC system coupled to a methanol reformer for military applications and is commercially available today. UltraCell's XX25 system, delivering 25 Wel, weighs only 1.24 kgs and is of the size of a video cassette, offering fuel capacity options for the XX25 that range from a compact 250 cm³ cartridge, which provides about nine hours of operation at a weight of about 350 grams, to a 18 litre tank that can provide uninterrupted power for days or weeks [113]. Within the scope of a BMBF (Ministry of education and research, Germany) funded project, experiments were performed with a mini HT PEMFC stack (containing BASF's Celtec P® MEAs), delivering 100 W_{el}, which was coupled to a methanol reformer developed by IMM, Mainz in Germany [114]. Vaillant, BASF and Plug Power have been working on a 5 kW CHP units (with a HT PEMFC stack coupled to a natural gas reformer) for residential applications [115].

Generalized fuel reforming steps:

Hydrogen, which is the preferred fuel for fuel cells in general, is not a primary energy, but it must be produced from other primary energy sources such as hydrocarbon based liquid or gaseous fuels or alcohols. For a desired pollution free environment, decarbonisation of fuels used in automobiles, mobile electronic devices, CHP units has been the most preferred approach. Therefore, hydrogen has been chosen as a preferred fuel of choice, though it is not a primary energy carrier. As hydrogen infrastructure is not yet realized, to operate fuel cell stacks to produce power, it has to be produced either from fossil fuel energy or from renewable energy. Renewable energy sources are the most preferred sources of energy due to their low carbon foot print compared to the fossil fuels. However, the overall efficiency (well to wheel) of hydrogen produced from renewable fuel (wind, solar, biomass, biogas, and the likes) has to compete with that produced from fossil fuels under various scenarios.

Hydrogen production from liquid or gaseous hydrocarbons and alcohols has been adopted by many fuel cell researchers to operate their systems based on fuel cell stack assemblies. On site production of hydrogen by steam reforming of natural gas, methanol or ethanol, propane or butane and subsequent feeding of hydrogen-rich gas into fuel cell stacks is a well known approach. Many groups have been working on various reformers (steam reformers, autothermal reformers, etc) to feed fuel cell systems of their choice, be it SOFC, MCFC, PEMFC, HT PEMFC. For instance, Thomas Aicher [116] has discussed various ways to couple fuel cell stacks as shown in Fig 3.35. Although not shown by Aicher, HT-PEMFC can be coupled to a reforming step as shown in Fig 3.35.

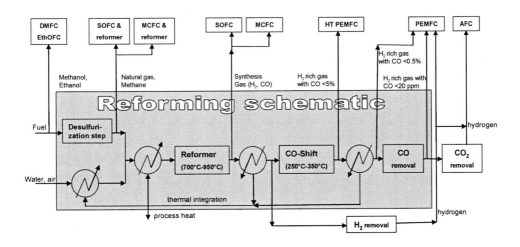

Figure 3.35: Possible ways to couple fuel cell stacks to reformers [116]

Hydrocarbons can be converted to produce hydrogen rich gas which follows a generalized process reaction shown in Eq 42 followed by water gas shift reaction (WGS) shown in Eq.43 to produce hydrogen.

Generalized equation for production of H_2 from HCs $$C_nH_m + nH_2O \longrightarrow nCO + (n+m/2)H_2$$	**Eq. 42**
Water Gas Shift Reaction Step $$CO + H_2O \longrightarrow CO_2 + H_2$$ $$(\Delta H°_{298} = -41.2 \text{ kJ/mol})$$	**Eq. 43**
Methanol decomposition at > 300°C or 573°K $$CH_3OH \longrightarrow 2H_2 + CO$$ $$(\Delta H°_{298} = +100.5 \text{ kJ/mol})$$	**Eq. 44**
Steam reforming of Methanol $$CH_3OH + H_2O \longrightarrow 3H_2 + CO_2$$ $$(\Delta H°_{298} = +59.3 \text{ kJ/mol})$$	**Eq. 45**
Steam reforming of Methane $$CH_4 + 2H_2O \longrightarrow CO_2 + 4H_2$$ $$(\Delta H°_{298} = +164.8 \text{ kJ/mol})$$	**Eq. 46**
Steam reforming of Bio-Ethanol $$C_2H_5OH + 3H_2O \longrightarrow 2CO_2 + 6H_2$$	**Eq. 47**
Steam reforming of Propane $$C_3H_8 + 6H_2O \longrightarrow 10H_2 + 3CO_2$$ $$(\Delta H°_{298} = +374.23 \text{ kJ/mol})$$	**Eq. 48**
Partial Oxidation (PrOx) of Methanol $$CH_3OH + 0.5 O_2 \longrightarrow 2H_2 + CO_2$$ $$(\Delta H°_{298} = -154.9 \text{ kJ/mol})$$	**Eq. 49**

Similarly, Eq.44 depicts methanol conversion at temperatures > 300°C (or 573.15°K). Eq.45 indicates steam reforming of methanol, which includes the WGS reaction (which is exothermic). Eq.46 depicts steam reforming of methane, which is endothermic.

The generalized reaction shown in Eq.42 is typically a strong endothermic reaction. But the overall heat of reaction normally depends on process conditions such as steam to carbon (S/C) ratios, temperature, pressure, etc. In an autothermal reactor, (or a partial oxidation step), air is often used and the reformate gas would have less hydrogen compared to the reformate from a steam reforming process [117]. Eq.49 shows methanol partial oxidation reaction which is strongly exothermic [118].

Ethanol, as it is also a renewable fuel, has emerged as a promising fuel to produce hydrogen rich gas. Eq. 47 depicts ethanol steam reforming reaction [119].

Furthermore, propane steam reforming can be explained by Eq.48. As propane is commercially available fuel, reforming of propane has been explored to produce hydrogen rich gas to be fed into fuel cell stacks [120], for their use in combined heat and power applications.

Results of simulations performed by Mark Saborni at the University of Duisburg-Essen, using Aspen ®, for steam reforming of methanol at steam to carbon (S/C) ratio of 1.5 for temperatures between 100°C- 500°C are depicted in Fig 3.36. As can be seen, operating in the temperature range of 240°C to 350°C, typical hydrogen concentration is around 75%, CO_2 is around 24% and CO concentration varies from 1% at 230°C to 4.3% at around 350°C. Methane and methanol concentrations are almost negligible.

With regard to propane being considered as a fuel for fuel cells, direct propane fuel cells (DPFC) are known for about 50 years where propane gas was fed directly into PAFCs operating between 150°C and 200°C, without reforming [121,122]. But the main challenge in those cells was that the anodic overvoltage was as high as 500 mV. This was attributed to the surface processes involved when propane was present on the catalyst surface [123] and the performance of such a cell was dependant also on the concentration of the electrolyte (PA). However, it was possible to operate PAFCs fed directly with propane, though not with high cell voltage. With regard to methanol as a fuel for HT PEMFCs, it was explained in section 3.2, how a Celtec V can be operated with methanol as a fuel being directly fed into a cell built with PBI/PVPA based membrane (or a high temperature DMFC).

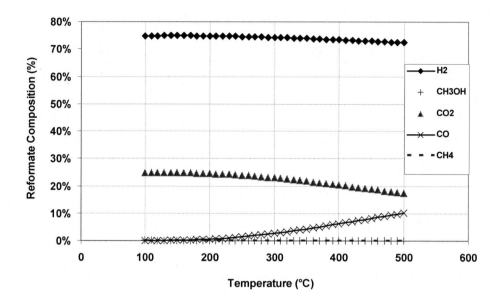

Figure 3.36: Simulated (Aspen ®) MeoH reformate composition (S/C: 1.5)

3.3.2 Influence of Natural gas, Propane and Methanol reformates

A HT PEMFC stack (Fig 3.37), was operated with three types of synthetic reformates. Methanol reformate, propane reformate and natural gas reformates with compositions shown in Table 3.17 were used as fuel gases and oxidant in the form of air was supplied.

Table 3.17: Composition of synthetic reformates fed into HT PEMFC stack

Reformate	Composition
Methanol steam reformate (S/C: 1.5) & reforming temperature of 350°C	H_2: 75%; CO: 1%; CO_2: 24%
Propane reformate (S/C: 3.4) & reforming temperature of 650°C	H_2: 54.8%; CO: 0.55%; CH_4: 1.38%; CO_2: 16.47%; H_2O: 26.78%
Natural Gas (ATR) reformate (O/C: 0.45) & reforming temperature of 650°C	H_2: 35%; CO: 0.2%; CO_2: 12%; N_2: 52.8%

Figure 3.37: 12-Cell HT PEMFC Stack tested with 3 types of reformates

Natural gas autothermal reformate was chosen as it might represent one of the worst case scenarios (consisting of lower hydrogen content in the reformate stream). Methanol steam reformate might be the best case scenario and Propane reformate might be in between the best case and the worst case.

Fig 3.37 depicts a 12-Cell HT PEMFC stack consisting of 12 individual HT PEMFCs having Celtec® P 1000 MEAs, with a total active area of about 600 cm². This is an air cooled short stack with a nomial power output of about 165 W_{el} at about 0.6 Volts / cell. Every alternate cell has cooling channels integrated into the graphite compound based bipolar half plates used. The 12-Cell stack was then tested systematically with hydrogen and air first and subsequently with the said three types of reformates at 165°C, with stoichiometries of 1.2 and 2.0 respectively for fuel and air. The performance of the 12-cell stack when fed with hydrogen and air, which is the reference case, is shown in Fig 3.38. As can be seen in the figure, the blue curve represents the total electrical power delivered by the HT PEMFC stack (12 cells) when fed with hydrogen and air, the light pink curve shows stack voltage.

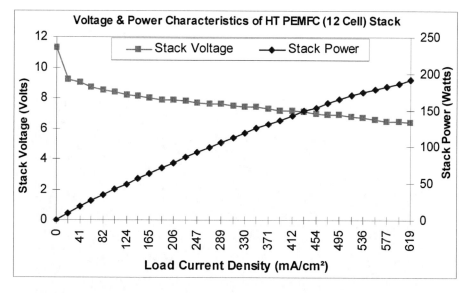

Figure 3.38: Performance of 12-Cell HT PEMFC stack fed with H_2 (λ = 1.2) & Air (λ = 2).

In Fig 3.39, the HT PEMFC stack performance when fed with the mentioned 3 types of reformates is compared to its reference case (when fed with hydrogen and air). The 12 cell stack was able to deliver about 78.4 Watts of electrical power at 200 mA/cm² of load current, about 112.4 Watts at 300 mA/cm² and about 166 Watts at 500 mA/cm², when operated at 165°C and with hydrogen and air. As can be seen from Fig 3.39, when fed with synthetic reformates, the power loss was more pronounced in case of natural gas autothermal reformate, when compared to the methanol steam reformate (which is commensurate with the H_2 content of the reformate). At higher load currents, this phenomenon was even more pronounced. For instance, the power loss with natural gas reformate at 200 mA/cm² load was 3.6 Watts, whereas the power loss with the same reformate at 500 mA/cm² was 20 Watts (pure H_2 and air feed at λ = 1.2 and 2.0 respectively being the reference case). Also, up to about 370 mA/cm² or until the voltage level per cell was around 0.6 V, the stack performance was not significantly affected by the reformate constituents (also because the CO concentration was below 1% in the reformate). The HT PEMFC stack exhaust gases from the anode were analysed using a gas analyser from Rosemount ®, Germany to check for the constituent gases of the reformates such as CO, CO_2 and H_2 [124]. From that study, it was observed that the CO which was present in the synthetic reformate being fed into the HT PEMFC stack had exited in its entirety from the anodic outlet. In conclusion, it can be said that, with the reformate feeds consisting more of inert gas composition (for instance N_2 and less of H_2), the dilution effect (lower H_2 partial pressure) was clearly seen from the lower stack performance.

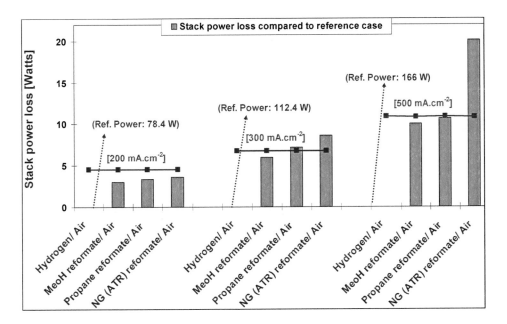

Figure 3.39: HT PEMFC stack performance comparison when fed with reformates

Furthermore, apart from CO tolerance and tolerance to other impurities, it is necessary that the individual cells of the HT PEMFC show stable and comparable performance. Individual cell voltages measured from the 12-cell stack are depicted in Fig 3.40, where the fuel used was synthetic methanol steam reformate. The individual cell voltages were observed to be stable over the entire load current range of 0 to 500 mA/cm². The lower individual cell voltages were observed in cells which were away from the cathode inlet. The flow field used in the individual bipolar plates was a 6 channel serpentine and the fuel and oxidant were fed in cross-flow pattern into these flow fields. Some of these lower cell voltages observed in individual fuel cells could be directly attributed to the patterns of oxidant supply. For instance, away from the cathode inlet, where air was fed, the individual cells may not have the very same amount of air supplied to them, compared to those near the cathode inlet. Furthermore, when load current is drawn, the produced water vapour (more pronounced towards the cathode exit) would offer a lot of resistance to oxygen diffusion into the active sites of the triple phase boundary where the ORR (oxygen reduction reaction) takes place. As explained in the previous sections dedicated to oxygen diffusion in HT PEMFCs, slow ORR in HT PEMFCs or PAFCs is a well known inherent problem which has to be taken into account. However, the individual cell voltages shown in Fig 3.40 demonstrated stable performance, possibly because issues related to water management are not present here. Whereas in the case of LT PEMFCs, deviation in individual cell voltages of a stack could mainly be attributed to non uniformity in membrane hydration level as well as in oxidant supply. Looking at Fig 3.40, the top blue dotted line is the HT PEMFC stack's performance when fed with pure H_2 and Air at 165°C.

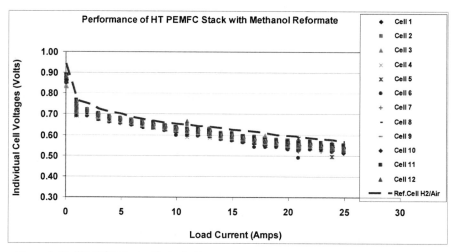

Figure 3.40: HT PEMFC stack's individual cell voltages when fed with methanol reformate

4 Long term operation of fuel cells

Durability of fuel cell stacks is a necessary pre-requisite before the on-set of their possible commercialization. Many groups have been working on this subject for a long time [125,126]. Many groups [127,128,129,130,131,175] have performed long term tests to study the performance of their MEAs, cell as a whole and their systems consisting of necessary balance-of-plant (BOP) components. LT PEMFCs, SOFCs, MCFCs, have been studied for their long term stability by some groups. Apart from long term operation on continuous load, load cycling, temperature and humidity cycling, no load operation, operating with freeze-thaw cycling [132] have been the focus of some investigations directed at fuel cells and their associated systems. This is because, various possible operating regimes of a fuel cell stack or a system have to be put to test under realistic operating conditions to draw useful conclusions, based on which commercial systems could be built. Some groups have recently focussed on BOP components, to understand their stability in greater detail. This chapter focuses on the subject of long term operation of fuel cells, with a special focus on HT PEMFCs.

4.1 General introduction to long term fuel cell operation

Testing of a fuel cell over longer periods (extending to thousands of hours of operation) needs specially designed testing protocols. In some cases, it may not be sensible to wait for the long term tests to conclude, and then design commercially viable systems. Therefore, accelerated ageing tests serve as an alternative methodology. Some aspects of fuel cell ageing, accelerated ageing are discussed in the following sections. However, the widely accepted long term testing procedure adopted by many fuel cell research groups (ZSW, DLR, FZJ) [133] is shown in Table 4.1.

Table 4.1: Testing protocol applicable to long-term testing of PEM fuel cells in general

Burn-in period (150 hours)	Long-term operation (with/without cycling) (1000s of hours)	Ex-Situ Tests
EIS, CV, I-V	EIS, CV, I-V	TEM, SEM, EDX

During the first 100 to 150 hours, usually a fuel cell's performance is analysed using EIS to study impedance contributions from cell components, CV to study the electrochemical surface area and I-V curves to study cell performance under various operating conditions such as cell temperature, fuel/oxidant stoichiometries, at various pressures and dew point temperatures. The first 100 to 150 hours is referred to as a burn-in period during which an electrochemical cell such as a PEMFC is conditioned. The same procedure is adopted by groups performing long term tests with HT PEM fuel cells [134,135]. During this period, it is expected that the catalyst sites, GDLs and the membrane find some equilibrium with regard to the fuel cell operating regime. For instance, a PEMFC fed with hydrogen and air at an operating temperature of 160°C, had shown very good performance at the beginning. But as the time goes by, the electrolyte finds its equilibrium with its surrounds, such as channels on the bipolar half plates, in coming gases, produced water, etc. Also, the catalyst sites which might have been dry at the start would get wet and further more, some catalyst sites might also get submerged in water which is produced in the cell. GDLs or MPLs (microporous layers) might also get filled partially with water vapour as the PEMFCs operate for some hours. The membrane material might creep, giving rise to its dimensional change, thereby affecting the impedance of the entire cell, as the fuel cell in question is operated for some hours. Similarly, at a particular load current, inlet gas stoichiometry and dew point temperature, the catalyst sites might selectively get closed. Therefore, it is believed that during the first 100 to 150 hours of cell operation at a reasonably rated current of about 200 mA/cm² of continuous load, the cell would find its balance with respect to its surrounds, etc. This period is known as a burn-in period. Understanding the impedance contribution from various cell components during this period is interesting to see how the cell impedance gets worse after many 1000s of cell operational hours. Similarly, understanding the total electrochemically active surface area of the MEA is interesting to note at the start of a planned long-term test, which could then be compared to the ECSA after the life test. The I-V curves at the beginning (BOL) and end of life (EOL) might help compare the overall performance of a PEMFC, which includes all the losses, processes, involved.

After the burn-in period, usually, the cell or stack in question is tested for many thousands of hours (as might be required by the user's group), with their specific testing protocol. For instance, an automobile company might be interested in seeing the performance of a FC stack when exposed to a load pattern which imitates a driving cycle. Also, a CHP unit supplier might be interested in the performance of a FC stack with its specific number of on/off load cycles, temperature cycling, etc. The long-term performance test is expected to give information about a FC stack's performance for a given situation for a given period of time. However as mentioned earlier, this is a time taking process, which may need to be replaced with an accelerated ageing test. However, it must be said that no accelerated ageing test can be identical to a real cell test. The following section briefly discusses the long term tests performed by some well known groups. The ex-situ test usually is performed at the end of long-term test, as this would mean disassembling of the fuel cell assembly.

4.1.1 LT PEMFC Versus HT PEMFC – Operation without humidification

The current author had performed a 1000 hour test with a commercially available MEA of 425 μm (30 μm thick membrane) with a Pt loading of 0.4 mg/cm² where hydrogen and air were fed at 1.2/2.0 stoichiometries without any active humidification. The water produced as part of cell reaction helped maintain membrane hydration at 40°C of cell operation, at a load of 200 mA/cm². The cell's performance loss rate was observed to be 33 μV/hr during the 1000 hour test (both in the case of a LT PEMFC and for the HT PEMFC) and from the LT PEMFC test, the sulfonic acid (SO_3^-) group loss rate was around 0.9 μg/hr (from the product water analysis), whereas the PA (phosphoric acid) loss rate from the HT PEMFC operation was around 0.25 μg/m²/s [175,136]. This test was intended to compare performance of a HT PEMFC single cell where fuel and oxidant are not humidified, with a LT PEMFC at comparable conditions. It was observed that LT PEMFC's degradation was higher and the cell voltage loss was rapid compared to the HT PEMFC, as operating LT PEMFCs with low or ultra low humidification limits their operation to low temperature and low load conditions, as the proton conductivity of the LT PEMFC based membranes is critically dependant on water.

4.2 Long term operation of HT PEMFCs

As mentioned earlier, HT PEMFC technology is rather new compared to the LT PEMFC technology. There are not many research groups or companies involved in long term testing of HT PEMFCs. BASF (Formerly known as PEMEAS) has reported long term performance of their HT PEM single cell with Celtec ® P 1000 MEA performed for > 18000 hours, with a voltage loss rate of about 6 μV/hr. They operated their cell at 160°C, with hydrogen and air with stoichiometries of 1.2 and 2.0 respectively and with a load of 200 mA/cm² [137,138]. It was also reported that under cyclic conditions (OCV – 200 mA/cm² – 600 mA/cm²), the degradation rate was around 27 μV/hr with a standard Celtec P 1000 MEA and with a MEA having a nanoparticle based filler, they observed a degradation rate of around 12 μV/hr. This implied a projected life of about 6000 hours without nanofillers and 14,600 hours with nanofillers. A 10% loss in cell voltage from its initial value at BOL (of around 0.65 V on 200 mA/cm² load) is taken by Schmidt et al as EOL [139]. They had also reported a cycling test (alternately operating between 200 mA/cm² at 180°C and shut down) with a MEA containing nano fillers as 0.13 mV per each cycle and the projected life was around 2000 hours under these conditions. Danish Technical University (DTU) has performed a single cell test with their own high temperature stable PBI/H_3PO_4 based MEA for about 5000 hours at 150°C with pure hydrogen and oxygen [140]. Forschungszentrum Jülich together with Fuma Tech, Germany had reported a voltage loss of around 25 µV/hr, based on their single cell operated on 200 mA/cm² load current for 1000 hours at 160°C [141].

4.2.1 Electrolyte loss during long term operation

One of the cell degradation mechanisms related to high temperature PEMFCs based on PBI/H_3PO_4 systems is electrolyte loss. Phosphoric acid could be leached out from an operating HT PEMFC depending on the operating parameters such as temperature, load current, inlet gas humidity, etc. Litt and Ameri et al [142,143] have elucidated that 2 moles of acid per each mole of PBI repeat unit typically is immobilized in these membranes. The remaining (X-2) moles of acid, which is referred to as excess acid or amorphous phase acid, contributes to proton conductivity in these HT PEMFCs. But part of this free, excess acid could be easily flushed out, which is very specific to cell operating conditions (such as dew point temperatures, conditions which facilitate higher water concentration gradients across the membrane, to name a few). Also, as H_3PO_4 is a hygroscopic acid, operating these HT PEMFCs at process conditions where liquid water is formed might also lead to acid leaching. In other words, acid retention is vital to ensuring long term satisfactory operation. Load cycling, temperature cycling, feed gas stoichiometries will also have their corresponding influence on acid retention. Evaporation of phosphoric acid from PAFCs was studied by Okae [144] in the 1990's. Whereas recently, Seel et al [145] have reported that PA evaporation in Celtec P® based membranes is about 1.5 – 2 times lower than those in PAFCs reported by Okae. Their (PAFCs') acid loss was dependant on acid concentration, temperature of the acid (electrolyte) and its surroundings (such as any ternary elements or gases). Recently, Scholta et al [146] have proposed that one of the approaches to handling acid evaporation in HT PEMFCs operating in the temperature regions of 150°C-170°C, while starting to operate them at temperatures as low as 60°C, is by supplying very high amounts of reactant gases (especially on the cathode side). However, this might be possible to realise when one or few cells are used in a HT PEMFC stack. But in bigger stacks, this may not be practical. Secondly, the supply of such high (10 to 100 times the stoichiometric feed) amounts of reactant gases may not be efficient, as it would mean an enormous BOP consumption. However, at temperatures above 180°C or 200°C, phosphoric acid might get dehydrated under certain conditions (such as low load and lower stoichiometric feed operation), as studied by Ma [147]. For all practical purposes, HT PEMFCs, to become viable, may have to offer satisfactory performance at temperatures of 150°C-180°C, with regard to acid (electrolyte) integrity. The vapour pressure of phosphoric acid was studied in much detail in the 1900s when developing fertilizers for soil enrichment. From the studies performed by Fontana [148], MacDonald [149], Earl H.Brown

[150], the vapour pressure of a closed P_2O_5-H_2O system is derived and shown in Fig 4.1. For increasing concentrations (while going from 85 wt% to 99 wt%)of PA, the vapour pressure decreases and as the temperature goes up, the vapour pressure also goes up. Whereas, the protonic conductivity of the electrolyte (PA) decreases with increasing concentrations and decreasing temperatures.

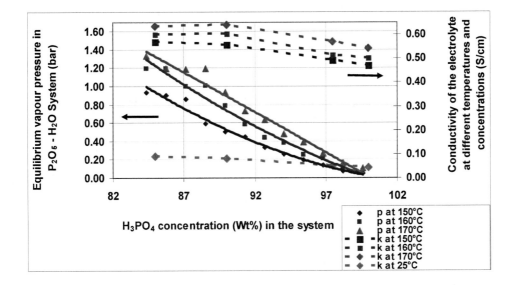

Figure 4.1: Vapour pressures and conductivities of the electrolyte 150°C–170°C. Green curve: 170°C; Brown: 160°C; Blue: 150°C; red: 25°C; p: pressure, k: conductivity.

For a highly doped MEA discussed in this work, 90 wt% of electrolyte concentration will be the most relevant, as enumerated by Schmidt et al [151]. For phosphoric acid of 90 wt% concentration, the water vapour pressure over phosphoric acid (Blue curve on Y-axis in Figure 4.1) will be 0.506 bar at 150°C, at 160°C it is 0.786 bar and at 170°C it is 0.932 bar. Therefore at higher operating temperatures, water evaporation from PA is a possibility, which results in lowering of electrolyte conductivity [152]. However, as the reactant gases flow over the electrolyte in an operating fuel cell, water evaporation from PA will be less of a concern. During start-up and shutdown phase of a HT PEMFC, acid retention can be achieved by filling the fuel cell with inert gases such as N_2 [153].

It can be recalled that, BASF's Celtec ® P based MEAs have very high amounts of acid (32 to 40 moles / mole of PBI repeat unit) in them, offering very high proton conductivity and subsequently high cell performance. Whereas the lowly doped MEAs such as the ones from

DTU, may have around 6.6 to a maximum of 10 moles / mole of PBI repeat unit, offering a slightly lower performance compared to the highly doped MEAs. However, Kerres and Li, et al [154] have argued that their membranes, (although not highly doped) do not lose much acid when compared to highly doped membranes and therefore, taking the long term operation of these HT PEMFCs into account, choosing a lowly doped membrane with much lower acid loss rates is preferred to highly doped ones with a higher acid loss rate. Celtec P® based MEAs were examined in the endurance test discussed in the following section.

4.2.2 Analysing HT PEMFC operation at 160°C and 170°C

Two single cells, each with an active area of about 50 cm² were operated for about 1000 hours each at 160°C and 170°C respectively to study their long term performance and acid loss. These two single cells were constructed with cell components similar to the ones depicted in Fig 3.1 and containing commercially available Celtec® P1000 from BASF, Germany. It can be seen in the following paragraphs that at higher operating temperatures of the HTPEM fuel cell, although the cell's performance is expected to be higher due to enhanced kinetics, its long-term performance (> 1000 hours) is inferior to its performance at lower temperatures. The higher electrolyte loss from the membrane was observed to be one of the main reasons for this cell degradation at higher temperatures.

With an initial phosphoric acid content of about 32 moles per mole of PBI repeat unit, these MEAs exhibit conductivities close to 0.25 S/cm [85] at around 160°C.

A test stand similar to the one shown in Fig 3.2 equipped with Lab View ® software and other hardware from National Instruments, Texas, USA was employed to monitor, control and log the parameters such as temperature, voltage, current, fuel and oxidant flow, through both of the HT PEMFCs examined here. MFC for methane was added to the anode side manifold of the test stand (Subsystem I) shown in figure 3.2. Each HT PEMFC (single cell) was heated using heating cartridges inserted into its endplates. Firstly, after heating the cell to about 170°C, fuel hydrogen and oxidant (air) were supplied at stoichiometries of 1.35 and 2.5 respectively, while drawing some load current. The current drawn was limited to 200 mA/cm², which is 10 amps in each case. While maintaining the cell temperature at the required level, the cell voltage was monitored. The PTFE pipes connected to anode and cathode outlets of the single cell (see Fig 3.25) were directed to go though a container where the product water was condensed. The product water thus collected was analysed using ICP-MS to ascertain the amount of phosphoric acid present in it, in each case. Each single cell was initially operated for about 100 hours, which was considered as a necessary conditioning period. Cell per-

formance after this initial conditioning period is considered for further analysis. The performance of the two single cells from their 1000 hour test is plotted against cell's operating hours
and is shown in Fig 4.2. The primary y-axis shows the observed cell voltage in each case
and the secondary y-axis shows the phosphoric acid loss (from product water) rate from each
cell over the tested period. Fig 4.3 shows the linearized single cell performance in both
(160°C and 170°C operation) cases over a 1000 hour testing period.

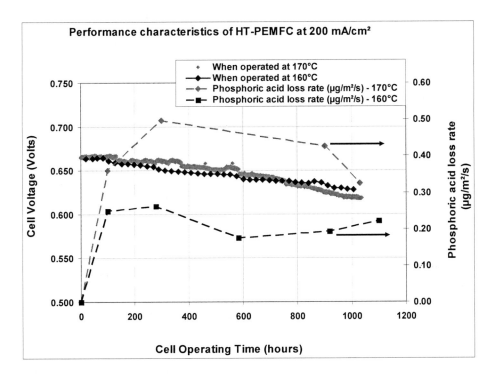

Figure 4.2: HT PEMFC's (single cell) performance and acid loss rate. Black curve: 160°C;
Grey curve: 170°C.

As depicted in the figures 4.2 and 4.3, the cell voltage was 0.665 Volts when operated at
170°C, on 200 mA/cm² load at BOL, whereas the same at 160°C was 0.661 Volts. It can be
seen that the cell operated at 170°C had exhibited higher cell voltage during the first 800
hours of cell operation. The initial 100 hours of conditioning period was not taken into account. And thereafter, its performance fell below the performance of the single cell operated
at 160°C.

It is also interesting to see that the phosphoric acid loss rate was higher in the cell operated at 170°C. It can be observed from Fig 4.2, that during the first 300 hours, the phosphoric acid loss rate was the highest and this trend was similar in both the cases. During the subsequent 700 hours of cell operation, the electrolyte (H_3PO_4) loss rate remained lower. The amount of phosphoric acid leached out from the MEA is expressed here as micrograms of PA per square meter of MEA per second of fuel cell operational period and is plotted in the secondary y-axis of Fig 4.2.

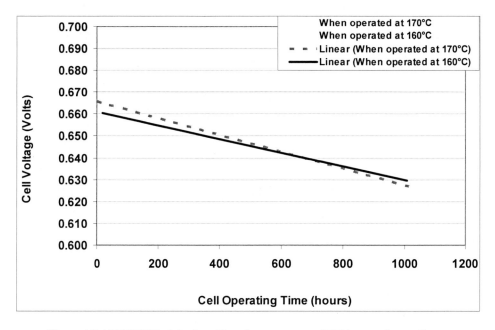

Figure 4.3: HT PEMFC's (single cell) performance over 1000 hours of operation

After 100 hours of cell operation at 170°C, the total amount of PA lost from the cell was 650 µg, whereas the total PA lost from the cell operated at 160°C, was 450 µg. After 900 hours, acid loss was 7,040 µg from the first cell (170°C) and it was 3,400 µg from the second cell (160°C). That is, after 900 cell operational hours, the total acid lost from the first cell (170°C) was double the amount lost from the second one. It was shown while discussing vapour pressures in P_2O_5-H_2O system (Fig 4.1), that going from 160°C to 170°C, in the case of a 90 wt% phosphoric acid, there would be a 18.5% rise in its vapour pressure. This would have two immediate consequences: (i) water evaporation leading to PA dehydration and lowering of electrolyte conductivity, (ii) water-acid evaporation. The former was discussed already, and had been studied extensively by Higgins et al in 1955 [155]. The later is not yet clearly understood, although the acid loss from fuel cell product water analysis (when a HT PEMFC is operated at higher temperatures) had confirmed it. Plug Power Inc. had observed similar phenomenon [156]. Recently Oono et al [157] had observed similar phenomenon while performing long term tests with their own MEA. Also, studies performed by Okae et al [158] observed PA loss as the operating temperature rises. After 900 hours, the first cell had lost 40 mV, whereas the second one had lost 29 mV. That is, the cell's performance loss rate was 44 µV/hr in the first case, whereas the same was 32 µV/hr in the second case. It can be concluded that the electrolyte loss in HT PEMFCs based on PBI/ H_3PO_4 systems can be higher at higher cell operating temperatures. In the current study, a 10°K rise in temperature resulted in an electrolyte loss rate of about 2.0 times its value at 160°C. However, it must be noted that the CO (Carbon Monoxide) tolerance of a HT PEMFC operating at 170°C will be higher compared to the cell operating at 160°C. For instance, a single cell operating at 170°C could tolerate 1% CO with a loss of about 15 mV in cell voltage at 200 mA/cm². Whereas the same at 160°C could lose about 25 mV, under same conditions, when Celtec® P 1000 MEAs are used in a HT PEMFC [159].

4.2.3   2000 hour test of a HT PEMFC with Celtec P 2000 MEA

Results pertaining to a single cell which was tested for about 2000 hours constructed with Celtec P® 2000 MEA from BASF, Germany are discussed in this chapter.

A single cell of the type discussed in chapter 3 was assembled with a Celtec P® 2000 MEA. The active area of the cell was 50 cm². The cell was operated with H_2/Air, with stoichiometries of 1.35/2.5 respectively at 160°C of cell temperature. This cell was operated for > 2000 hours on a load of 200 mA/cm² (10 amperes). Fig 4.5 depicts the observed performance of this single cell HT PEMFC. Fig 4.4 depicts the test stand used.

Figure 4.4: Test stand where HTPEM single cell was tested for 2000 hours

The X-axis in Fig 4.5 indicates the cell operating hours, Y-axis indicates the cell voltage in volts. The black curve indicates daily average voltages. The performance curve is divided into zones 1 to 7 to analyse and discuss the test results. However, the performance of a single cell operating on a typical working day at the set load of 10 amps and at 160°C, with hydrogen and air stoichiometries of 1.35 and 2.5 can be seen from Fig 4.6. The X-axis denotes the time during which the single cell was operating (from 00:00 hrs in the morning till 00:00 hrs on the next morning). The primary Y-axis shows the cell voltage in volts, air feed in litres/min, hydrogen feed in litres/min. The secondary Y-axis shows the cell temperature.

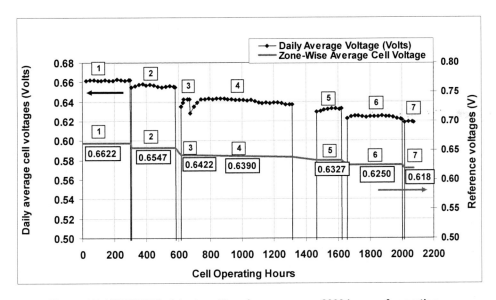

Figure 4.5: HT PEMFC's (single cell) performance over 2000 hours of operation

Table 4.2: Breakup of Cell Voltage loss – Zone wise

Zones	Events	Hours	Voltage loss	Actual Cell Voltage	Average Cell Voltage
		(hr)	mV	Volts	Volts
Start	Constant load : 200 mA/cm²	0	-	0.6610	-
1	Constant load : 200 mA/cm²	300	-1.3	0.6623	0.6622
1 - 2	Load cycling : 0 - 400 mA/cm² - 2 times	6	7.3	0.6550	-
2	Constant load : 200 mA/cm²	273	0.3	0.6547	0.6547
2 -3	Cell idling (including on OCV: 20 mins, no resistor across Cell)	39	19.9	0.6348	-
3	Constant load : 200 mA/cm²	49	-7.4	0.6422	0.6422
3 - 4	Cell Idling (160°C)	6	-	-	-
4	Constant load : 200 mA/cm²	645	5.1	0.6371	0.6390
4 - 5	Cell idling	152	7.6	0.6295	-
5	Constant load : 200 mA/cm²	156	-3.4	0.6329	0.6327
5 - 6	Cell idling (including on OCV: 2 mins; Resistor across Cell)	36	10.3	0.6226	-
6	Constant load : 200 mA/cm²	345	1.2	0.6214	0.6250
6 - 7	Load cycling 6 h + Cell idling (including on OCV: 2 mins, then Resistor across Cell)	15	3.0	0.6184	-
7	Constant load : 200 mA/cm²	59	-0.4	0.6188	0.6188

111

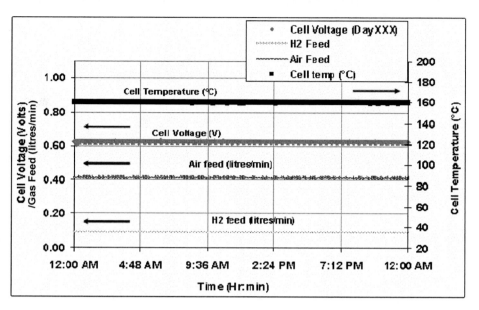

Figure 4.6: HT PEMFC's (single cell) performance during 24 hours of operation

The HT PEMFC's performance degradation was studied by dividing the cell's performance curve into seven zones, which are shown using square boxes above the curves in Fig 4.5. Each zone from 1 to 7 represent HT PEM single cell's performance at a constant load of 10 amperes (200 mA/cm²) which are separated by 5 types of test interruptions. The observed cell voltage was 0.6610 Volts at the start of the life test at BOL (after first 100 hours of cell operation on 10 amps). Then, the test was a continuous load test (200 mA/cm² at 160°C) in zone 1, for 300 hours, after which the load was cycled between 0 and 400 mA/cm² at 160°C. This operation lasted 6 hours, after which the load was set to 200 mA/cm². The dark points show daily average cell voltages with the respective zone numbers shown above them. The grey lines show the average cell voltage in each zone.

Details are shown event wise, in Table 4.2. The first column in Table 4.2 shows events 1 -7 and the events between zones such as 1-2, 2-3, 4-5, 5-6, 6-7. The second column shows the event or the mode of cell operation. The third column shows the operational hours in that given mode of operation. The fourth column shows the cell voltage loss seen by the cell in each mode, corresponding to its mode of operation. The firth column shows the actual measured cell voltage that was recorded at the beginning of each mode. The last column shows the average cell voltage in each mode. These values can also be seen in Fig 4.5.

1) In Zone 1, it was constant load operation on 200 mA/cm² load at 160°C for 300 hours. The cell had gained about 1.3 milli Volts (See Fig 4.5).

2) Between Zone 1 and Zone 2, the cell voltage loss of 7.3 mV, the test event was load cycling from 0 – 400 mA/cm², in steps of 40 mA/cm², for 6 hours.

3) In Zone 2, it was constant load operation on 200 mA/cm² load at 160°C for 273 hours and the voltage loss was 0.3 mV during the 273 hour operation. This corresponds to a voltage loss of 1.09 µV/hr.

4) Between Zone 2 and Zone 3, the cell was operating on no load for 20 minutes at the same operating temperature of 160°C. After this 20 minutes, the cell Voltage fell to 0 Volts in 10 minutes, as anode was fed with nitrogen and the cathode was fed with air and the cell was kept in idling mode (after disconnecting the load, closing in and out-lets of the cell) at room temperature. The cell had lost 19.9 mV during these 39 hours in this phase.

5) In Zone 3, it was constant load operation on 200 mA/cm² load at 160°C for 49 hours and the cell had gained 7.4 mV during the 49 hour operation.

6) Between Zone 3 and Zone 4, the cell was fed with nitrogen on the anode side and air on the cathode side and was kept at 160°C. The flow rates were 500/500 ml/min re-spectively. Although there was an initial fall in cell voltage (3.2 mV) in this phase, the cell voltage improved continuously in Zone 4.

7) In Zone 4, it was constant load operation on 200 mA/cm² load at 160°C for 645 hours and the cell had lost 5.1 mV during the 645 hour operation, pointing to a voltage loss rate of 7.9 µ/hr.

8) Between Zone 4 and Zone 5, the load was disconnected; cell was cooled down to room temperature by feeding nitrogen and air of 500/500 ml/min respectively. The cell was kept in this mode for 152 hours. The observed voltage loss was 7.6 mV.

9) In Zone 5, it was constant load operation on 200 mA/cm² load at 160°C for 156 hours and the cell had gained 3.4 mV during the 156 hour operation.

10) Between Zone 5 and Zone 6, the cell was on OCV for 2 mins and 88 seconds (on more than 0.8 V). But this time, a small 1 kΩ resistor was connected across the terminals of the HT PEMFC single cell, immediately when the electronic load was cut off to bring down the cell voltage quickly, by activating an electro-mechanical relay controlled by the test stand hardware. The cell's in and outlets were closed and the cell was in this mode for 36 hours. The observed cell voltage loss was 10.3 mV.

11) In Zone 6, it was constant load operation on 200 mA/cm² load at 160°C for 345 hours and the cell had lost 1.2 mV during the 345 hour operation, pointing to a voltage loss rate of 3.4 μV/hr.

12) Between Zone 6 and Zone 7, load cycling was performed (0 to 200 mA/cm²) to examine cell's performance, at 160°C which lasted 6 hours after which the cell was on OCV for 2 mins and 42 seconds (on more than 0.8 V of cell voltage). This time also, the resistor was connected to the cell, the moment the electronic load was switched off by the LabView® program. After this, the cell voltage fell to 0 Volts when fed with nitrogen and air at flow rates of 500/500 ml/min respectively after which the cell's in and outlets were closed. The cell was in this mode for 9 hours. In total, between Zones 6 and 7, it was 15 hours. The cell's voltage loss was 3 mV.

13) In Zone 7, it was constant load operation on 200 mA/cm² load at 160°C for 59 hours and the cell had gained 0.4 mV during the 59 hour operation.

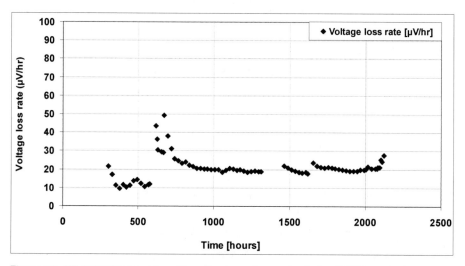

Figure 4.7: HT PEMFC's (single cell) performance loss rate during 2000 hour test

To summarize, Fig 4.7 depicts the cell voltage loss rate (loss in cell voltage/cell operational hours). Looking at Fig 4.7, during the initial 600 hours, the degradation rate was around 10 μV/hr, after which the massive cell voltage loss was encountered (due to OCV operation for about 20 minutes). After this voltage loss, the cell voltage loss rate from about 850 hours to 2000 hours of cell operation was on an average, 20 μV/hr. The operational phase with the lowest degradation rate was 7.9 μV/hr in zone 4. Start stop and cell idling, which includes load and temperature cycling (between each test start up) coupled with OCV operation has clearly induced cell voltage degradation.

Fig 4.8 shows the cell's current voltage curves at BOL and EOL when cycled between 20 and 400 mA/cm² at regular intervals. The dark red lines in Fig 4.8 show cell voltages at correpsonding load currents shown as orange points. The green points show cell voltages at corresponding load currents shown as blue points. The latter refer to cell's performance at EOL and the former to BOL. As the difference between these two curves can hardly be seen from Fig 4.8, Fig 4.9 is used to highlight the differences in cell voltages. Fig 4.9 shows linearized cell performance on 20 to 400 mA/cm² of load current at 160°C, being fed with hydrogen and air at stoichiometries of 1.35/2.5 respectively at BOL and EOL. (EOL in this work represents a performance loss of 6.6% from BOL). At BOL, on 20 mA/cm² load, the cell voltage was 0.8003 V and at the EOL, at the same current, the cell voltage was 0.7887 V (See Fig 4.8).

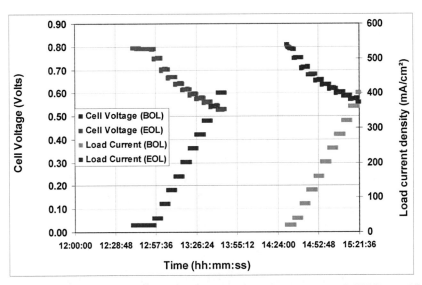

Figure 4.8: Performance of the single cell at BOL and EOL: Dark red data: Cell Voltage at BOL, Orange points: Load at BOL, Green data: Cell voltage at EOL, Blue data: Load at EOL.

Furthermore, from the polynomial equations representing BOL and EOL from Fig 4.9, it can be said that the cell had lost about 2.3 mV/amp from its initial cell potential at BOL in the 1-20 ampere current range (for e.g., a total of 23 mV at 10 amps) at EOL. Although Celtec P® 2000 based MEAs are expected to be more stable compared to Celtec P® 1000 based ones, the observed performance loss was 10 µV/hr on constant load and an average of 20 µV/hr (with constant load, load cycling and temperature cycling, shutdown and start-up phases included) for the said 2000 hours of cell operation. During the course of the long term test, the product water collected from the cathode and anode outlets and was subsequently analysed (using ICP-MS technique) to ascertain the amount of phosphoric acid present in it. The other elements present in the water samples are shown in Fig 4.10.

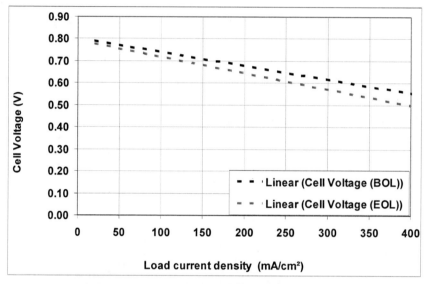

Figure 4.9: Linearized Single cell I-V Curves at BOL and EOL (Black: BOL; Grey: EOL).

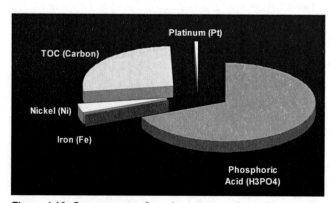

Figure 4.10: Components of product water collected (weight%)

It can be seen from Fig 4.10 that the major constituent of the product water was phosphoric acid, followed by particles of carbon. It was not clear whether they were from the GDEs of the MEA or from the graphite compound based bipolar half plates contained within the cell assembly.

Table 4.3: Constituent elements present in the product water collected from HT PEMFC

Items	Mesurement standards	Sample 1 mg/litre	Sample 2 mg/litre	Sample 3 mg/litre	Sample 4 mg/litre	Sample 5 mg/litre	Sample 6 mg/litre
Phosphorous (P)	DIN EN ISO 11885	1.600	0.72	0.69	0.68	0.67	1.1
Phosphate (PO_4^{3-})	DIN EN ISO 11885	4.900	2.2	2.1	2.1	2	3.3
Phosphoric Acid (H_3PO_4)	DIN EN ISO 11885	5.062	2.278	2.183	2.151	2.120	3.480
Iron (Fe)	DIN EN ISO 11885	0.026	< 0.01	< 0.010	< 0.010	< 0.010	< 0.010
Nickel (Ni)	DIN EN ISO 11885	0.260	0.1	0.096	0.081	0.08	0.071
TOC (Carbon)	DIN EN 1484	2.000	1.6	1.4	1.4	1.3	1
Platinum (Pt)	DIN EN ISO 17294-2	0.036	0.026	0.022	0.021	0.021	0.015
Conductivity	in µS	12.9	6.4	6	6.4	5.8	8.7
Acidity	as pH	5.00	5	5	5	5	5

As may be noticed from Table 4.3, except for the first and the 6th sample, the rest of the 4 samples have similar phosphoric acid content. The first sample was taken after the initial conditioning period and cell operation of about 200 hours. The samples 2 to 4 are during the constant load operation and sample 6 during load and temperature cycling. It can be seen that during cyclic operation, there was more acid (electrolyte) loss from the cell. However, an amount of about 3.48 mg of phosphoric acid per litre of water collected (over 297 hours) corresponds to 0.65 µg/m²/s of PA loss. Also, a loss of 2.27 mg of phosphoric acid per litre of water collected corresponds to the electrolyte loss of 0.42 µg/m²/s. It was estimated by Schmidt et al from BASF that acid loss of 0.5 µg/m²/s implies that the corresponding membrane could operate for about 40,000 hours before it could reach the end of life, which could be defined as the time taken to lose 10% of its initial performance (cell voltage) measured at the beginning of life. These calculations were based on the total acid contained in a Celtec P based MEA before cell assembly. In conclusion, it may be said that electrolyte loss during start-stop cycles of a HT PEMFC could reduce its durability if the start stop operations do not take care of acid management efficiently. This can be achieved by closing the in and outlets of the HT PEMFC during heating up phase and intermittent shutting off of the in and outlets during cooling down phase, to flush out any condensed water. While a HT PEMFC is cooled down from its 170°C to room temperature, water condensation takes place inside the connected pipes which must be flushed out, to avoid electrolyte loss. Also, the outlets of both anode and cathode of the cell could be directed through a container filled partially with ceramic pellets (or a condenser with a diaphragm based separator) by means of which, the liquid water collected in the bottle will not be able to re-enter the cell during the start-up phase.

4.2.4 Analysis of the electrochemical surface area (ECSA)

In this section, the electrochemically active surface area of the Celtec P® 2000 MEA before and after the 2000 hour test described above is analysed using cyclic voltammetry (CV) technique. Both anode and cathode electrodes were studied employing the procedure described here:

The electrode of interest or the working electrode (WE) was fed with 100 ml/min of nitrogen and the counter electrode (CE) was fed with 100 ml/min of hydrogen. Zahnermesstechnik's IM6 was connected to the fuel cell, in a two-electrode configuration, where the WE was connected to IM6's positive terminal and the CE to the negative terminal. Firstly, the cathode of

$$q_{pt} = \left[\frac{H}{vA} \right]$$

Eq. 50

Where,

q_{pt} represents charge density in Coloumbs/cm²

'H' represents area under H-adsorption curve (Volts · Amps)

'v' represents sweep rate' in Volts/sec

'A' represents catalyst layer's geometrical area in cm²

fresh Celtec P® 2000 MEA was considered as WE and 6 CV scans were performed with a sweep rate of 50 mV/s in the potential range 50 mV to 1000 mV versus the CE. From these 6 CV scans, the charge density q_{pt} due to H-adsorption during the reverse scan (shaded area in Fig 4.11) was determined, following the relation shown in Eq 50. These H-adsorption areas from Curves 1 to 3 in Fig 4.11 overlap each other (red, blue and green curves).

The black curve shows the entire CV scan (also for Curves 1 to 3). The non-faradaic part that arises due to charging and discharging of electric double layer capacitance of elec-trode/electrolyte interface, indicated by the blue-shaded region in Fig 4.11 was excluded from q_{pt}, as this part contributes to charge accumulation but not to charge transfer.

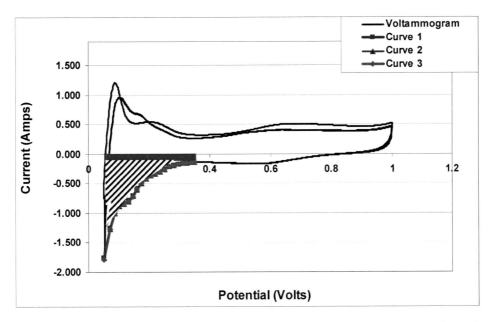

Figure 4.11: CV scan from a fresh Celtec P 2000 MEA's Cathode side: Curve 1 to Curve 3 show H-adsorption areas, performed 3 times. Black curve: CV scan over the en-tire range.

Electrochemically active surface area (ECSA) (cm²pt/cm²) =

$$ECSA_{pt} = \left[\frac{q_{pt}}{\Gamma} \right]$$

Eq. 51

where

q_{pt} is the measured charge density in C/cm².

$\Gamma = 210\ \mu C/cm^2\text{-pt}$ is a constant.

Similarly, 6 CV scans were performed on the fresh MEA's anode side. Also, after the 2000 hour test, CV scans were performed on the cathode as well as on the anode side of the used MEA. The results are summarized in Table 4.4. The determined charge densities were normalized to (Γ = 210 µC/cm²-pt), where Γ stands for charge required to reduce a monolayer of protons on Pt [160,161]. The normalized ECSA of different samples are shown in Table 4.4. ECSA values were further normalized to the Pt loading (0.4 mg/cm²) of the MEA electrodes and are shown in Table 4.4.

Table 4.4: Measured ECSA of Celtec P® 2000 MEA before and after 2000 hour test

SI No.	Description	Abbr.	Charge density (q_{pt}) C/cm²	ECSA cm²/cm²	Normalized Pt area cm²/mg-pt	Normalized Pt area m²/g-pt
1	New MEA (Anode)	1FA	0.047874	227.97	569.93	56.99
2	New MEA (Anode)	2FA	0.047579	226.57	566.42	56.64
3	New MEA (Anode)	3FA	0.047119	224.38	560.94	56.09
4	New MEA (Anode)	4FA	0.048438	230.66	576.64	57.66
5	New MEA (Anode)	5FA	0.048602	231.44	578.60	57.86
6	New MEA (Anode)	6FA	0.048230	229.67	574.16	57.42
7	New MEA (Cathode)	1FC	0.046872	223.20	558.00	55.80
8	New MEA (Cathode)	2FC	0.047242	224.96	562.41	56.24
9	New MEA (Cathode)	3FC	0.047221	224.86	562.15	56.21
10	New MEA (Cathode)	4FC	0.047014	223.88	559.69	55.97
11	New MEA (Cathode)	5FC	0.047337	225.42	563.54	56.35
12	New MEA (Cathode)	6FC	0.047349	225.47	563.68	56.37
13	(MEA-Life test-Anode)	1LA	0.047083	224.20	560.51	56.05
14	(MEA-Life test-Anode)	2LA	0.048223	229.63	574.08	57.41
15	(MEA-Life test-Anode)	3LA	0.048522	231.06	577.65	57.76
16	(MEA-Life test-Anode)	4LA	0.046776	222.74	556.86	55.69
17	(MEA-Life test-Anode)	5LA	0.047832	227.77	569.43	56.94
18	(MEA-Life test-Anode)	6LA	0.048122	229.15	572.88	57.29
19	(MEA-Life test-Cathode)	1LC	0.037781	179.91	449.77	44.98
20	(MEA-Life test-Cathode)	2LC	0.037883	180.39	450.98	45.10
21	(MEA-Life test-Cathode)	3LC	0.037783	179.92	449.80	44.98
22	(MEA-Life test-Cathode)	4LC	0.037855	180.26	450.65	45.07
23	(MEA-Life test-Cathode)	5LC	0.037893	180.44	451.11	45.11
24	(MEA-Life test-Cathode)	6LC	0.037808	180.04	450.09	45.01

Fig 4.12 compares the ECSA of different samples discussed. The abbreviations F-A stands for fresh MEA – anode side; F-C stands for fresh MEA – cathode side; L-A stands for life tested MEA – anode side; L-C stands for life tested MEA – cathode side. As is evident from the figure, there was no significant change in the anode side ECSA after the 2000 hour test, whereas on the cathode side, the ECSA before the long term test was 225 cm²/cm² and after the 2000 hour test, it was reduced to 180 cm²/cm², which is a 20% reduction in ECSA. It may be recalled that the cell had lost 42.2 mV during those 2000 hours, which is a 6.5% reduction in cell power.

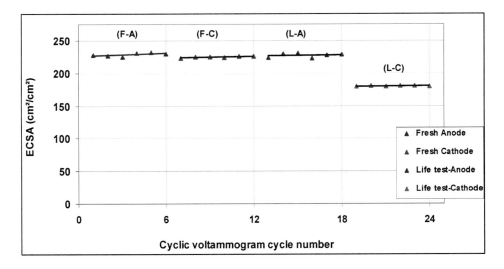

Figure 4.12: ECSA of samples before and after 2000 hour test. F-A: Fresh anode; F-C: Fresh cathode; L-A: anode after life test; L-C: cathode after life test.

In conclusion, (see Fig 4.12), it may be said that the cathode side ECSA is more prone to degradation as was also reported by some groups [162,163,164].

5   Degradation issues in Fuel cells - HT PEMFC Component challenges

Performance loss of fuel cells (or FC stacks) over time, which is also known as fuel cell degradation is a specialized field of research, which focuses on understanding the behavioural patterns of various cell components over time, degradation processes, in various fuel cell operating regimes. Some aspects related to fuel cell degradation in general and some specific degradation phenomena relevant to HT PEMFCs in particular, are discussed with necessary test results in the current chapter.

5.1.1 Fuel Cell degradation

Fuel cell degradation could be attributed to the collective degradation of its components. The essential components of a fuel cell such as a) Bipolar plates, b) Membrane, c) GDL or GDE d) Catalyst, e) Catalyst support, f) Microporous layer if any, g) Gaskets h) Current collectors, and other BOP components might degrade, with their characteristic degradation rate, dictated typically, by the type of material (and the corresponding properties of those materials) used in cell assembly, also depending on the operating parameters of the fuel cell considered. Extensive research has been performed over the past four decades to better understand various degradation phenomena related to fuel cell components such as membrane degradation and catalyst degradation of LT PEMFCs. Although some useful conclusions could be drawn from these experiences, issues related to HT PEM fuel cell components are unique in nature and are specific to its operating environment such as higher operating temperatures, phosphoric acid environment, as enumerated further.

5.1.2 Bipolar plate related degradation

Bipolar plates (or a combination of two bipolar half plates) have the task of separating reactant gases (e.g., hydrogen and air) and also transferring current from GDL to the current collectors where load terminals are located. Also, bipolar plates have to be mechanically strong, so as to withstand any vibrations or any other mechanical impact stresses during fuel cell operation. When cooling channels (to take away the heat produced in a FC stack) are made on the second side of a bipolar half plate, they should be impermeable to reactant gases such as hydrogen and oxygen. Furthermore, sometimes, these bipolar half plates are subject to oxidizing conditions on one side and reducing conditions on the other side. Therefore, bipolar plates have to be chemically, electrochemically and mechanically stable over prolonged periods of time.

Metallic bipolar plates have the problem of surface deterioration (or corrosion) when used in fuel cell environments. Many groups [165,166,167,168] are therefore adopting graphite-compound based bipolar half plates, which usually consist of graphite, a polymer binder and a ternary compound such as carbon and sometimes carbon nanotubes. Although these graphite compound based plates might offer better electrochemical stability compared to metallic bipolar plates, their conductivites are much lower compared to metallic ones and their mechanical strength is some orders of magnitude lower than their metallic counterparts [169,170,171,172]. The manufacturing cost of these graphite compound based BPPs also is some orders of magnitude higher than the metallic ones.

In the event of any BPP based surface deterioration, during the long term operation of a HT PEMFC, the contact resistance between its GDL and its bipolar plate surface might increase over time, which would lead to performance deterioration. Some tests performed with bipolar plates used in a HT PEMFC are discussed here.

The bipolar half plate shown in Fig 5.1 is a graphite compound based plate meant to be used in HT PEMFCs and was developed by ZBT GmbH [173,174]. The polymer binder used in this plate was PPS (poly phenelyne sulphide).

Figure 5.1: PPS based bipolar half plate used in the long term performance test.

A single HT PEM fuel cell consisting two bipolar half plates of this kind (with a PPS binder), containing a Celtec® P 1000 MEA from BASF was used to perform a test of about 3000 hours. This single HT PEMFC was operated at 160°C at a load current of 200 mA/cm². The fuel gases supplied were hydrogen and air (λ = 1.35 and λ = 2.5 respectively). After the life test was concluded, the bipolar plates were analysed using SEM (Scanning electron microscopy) to examine any surface deterioration that might have been caused due potentials on the cathode and anode side of the BPPs, acid (PA) impact on the polymer, deposition of any other materials (for instance from the gasket material, coatings on the current collector plate). The SEM images from the bipolar half plate used on the cathode side are shown in Fig 5.2 and Fig 5.3.

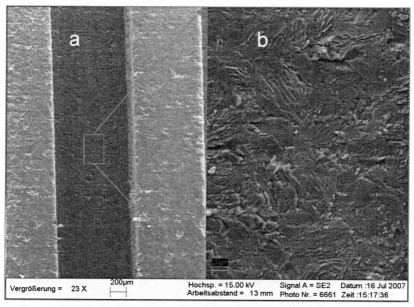

Figure 5.2: a) Surface of the PPS based bipolar half plate (Cathode side) after 3000 hour in - cell test at 160°C. b) Centre of flow-field channel.

A careful examination of the surface of the bipolar half plates at various points across their entire area has revealed that these plates were stable after the long term operation. However, phosphoric acid was observed on the surface of the bipolar half plates examined using SEM, at the university of Duisburg-Essen, Duisburg, Germany. Fig 5.3 depicts one part of the bipolar plate surface where deposits of PA were observed. It is likely that the excess acid which could have been squeezed out of the highly doped MEA, at the beginning of life (BOL) could have deposited itself on the surface of the bipolar half plate. Further, the electrolyte

124

(PA) that was leached out of the HT MEA, at the rate of about 0.2 µg/m²s (see Fig 4.2) could have formed a thin phosphate layer on the surface as was elucidated by the same author in the work presented at the Grove Fuel Cell Symposium, London, in 2007 [175].

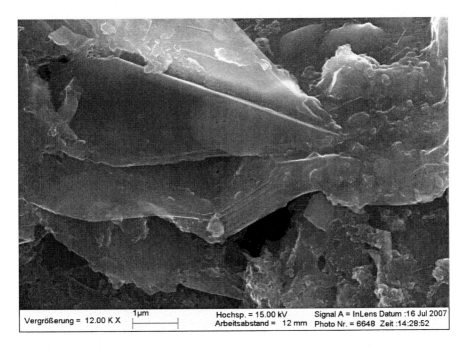

Figure 5.3: Deposits of phosphate observed on the surface of the PPS based
bipolar half plate (Cathode side) after 3000 hour in-cell test at 160°C.

In another work, one sample of these PPS based bipolar half plates were exposed to accelerated ageing test. This test was performed ex-situ where the high temperature stable bipolar half plate was immersed in 85% concentrated phosphoric acid and oxygen was supplied continuously and the container was heated to about 160°C. Fig 5.4 depicts the surface the bipolar half plate before the mentioned accelerate ageing test and Fig 5.5 depicts its surface after a 2000 hour ageing test. A careful examination of the bipolar half plate surface at different points under SEM had shown that there was no surface deterioration of the PPS based graphite bipolar half plate. But deposits of phosphoric acid were found on the surface of the plate as can be distinguished from the Fig 5.5. As this test was performed in a glass container, it was also observed that some silicon material (possibly from the glass container) was deposited on the plate surface. It is beyond the scope of this work to go into more details concerning the accelerated ageing test. However, some more details were discussed in the work presented by Derieth, et al [176].

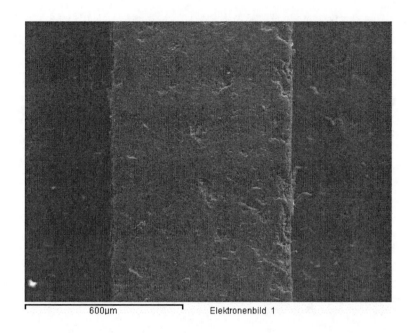

600µm   Elektronenbild 1

Figure 5.4: Surface of the PPS based bipolar half plate before accelerated ageing test.

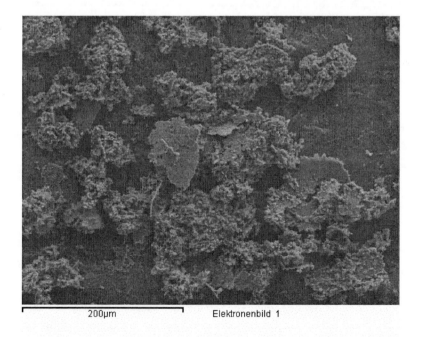

200µm   Elektronenbild 1

Figure 5.5: Surface of the PPS based bipolar half plate after accelerated ageing test.

5.1.3 Catalyst and Catalyst support degradation

Catalyst degradation is a well known phenomenon in PEMFCs. Catalyst related issues need to be adequately addressed when it comes to HT PEMFCs. Ferreira and Gasteiger [163] have elucidated that platinum dissolution increases with temperature in LT PEMFCs. Stability of platinum catalysts was found to decrease with increasing cell temperature as reported by Stevens and Wang [177,178]. Although not much was reported concerning catalysts for HT PEMFCs, some efforts are definitely underway to improve catalyst stability in HT PEMFC related MEAs [179]. However, most of the existing literature deals with ageing of platinum and platinum based alloys used in PAFCs. Also, catalyst materials which are suitable to be used in PAFCs are also suitable for use in HT PEMFCs. As PAFCs have been studied extensively in the past, much could be learned from the many years of research. Many groups have reported platinum particle growth in the cathode catalyst surface of a PAFC operating at 200°C [180,181,182,183,184,185]. Some of the mechanisms reported by Ferreira et al, based on their experiences with PAFCs were, a) Platinum dissolution and re-deposition caused due to Ostwald ripening, b) Pt particle agglomeration triggered by corrosion of carbon support (or when Pt particles detach themselves from carbon support), c) Coalescence of platinum nanoparticles via platinum nanocrystallite migration on the carbon based catalyst support. Some routes have been suggested to overcome this problem. For instance, presence of non-precious metal would enhance the stability of Pt catalyst as reported by Mukerjee et al [186], for use in LT PEMFCs. Combinations such as PtCo, PtNi on carbon support were observed to show better stability than Pt alone on carbon. In summary, binaries of type $Pt_xM_{(1-x)}$ where M stands for a secondary metal such as Ti, Cr, Mn, V, Fe, Co and Ni and x being in the range of 0.5 - 0.75 [187], for use in LT PEMFCs were explored as alternative catalysts.

With regard to catalyst support, carbon based supports experience severe degradation at increased cell potentials such as 0.8 V - 1.4 V range [188,189,190]. Fuel starvation on the anode side might lead to higher cell potentials where water can get oxidized [191,192]. Carbon corrosion reaction is shown in Eq.52. Catalyst support degradation results in catalyst detachment and thus results in cell performance loss.

BASF, Germany has studied the stability of their Celtec P® 1000 MEAs consisting highly doped PA based membranes and Vulcan XC 72 supported Pt and Pt alloy catalysts used for the anode and the cathode respectively [193]. At 180°C, they had observed a degradation rate of 26 μV/hr from their single cell tests using these MEAs. BASF also has reported the degradation of their Celtec P®1000 based catalysts under cyclic conditions and have reported a better performance of their catalysts used in their next generation MEAs called as Celtec P® 2000, as shown in Fig 5.6 [194]. Fig 5.6 depicts the cell voltage loss at 800 mA/cm² of load current, when their Celtec P® based MEAs were operated in cycling mode (going from 0.6 V to 0.8 V alternately). They had reported that their new carbon black materials used on the cathode side (Celtec P® 2XXX) have resulted in this improved stability. As can be noticed, cell characterization and cell durability studies reported in this work were also based on these Celtec P® 2000 based MEAs.

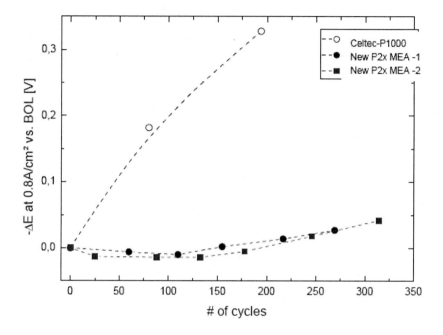

Figure 5.6: Peformance loss of cells containing Celtec P® based MEAs with cell Voltage cycling from 0.6 V to 0.8 V (adapted from BASF fuel cells, 2009)

$$C + 2H_2O \longrightarrow CO_2 + 4H^+ + 4e^- \quad \text{(Where } E° = 0.207 \text{ V)}$$

$$C + H_2O \longrightarrow CO_2 + 2H^+ + 2e^- \quad \text{(Where } E° = 0.518 \text{ V)}$$

Eq. 52

Therefore, operating HT PEMFCs at cell voltages close to 0.8 V must be avoided to avoid catalyst and catalyst support based cell degradation. In the current contribution, a 120 hour test was performed to study the cell performance loss when operated at open circuit voltage (or under no load conditions) and is discussed in the following paragraphs.

5.1.4 Cell degradation under OCV conditions

A single cell with a Celtec P® 2000 MEA from BASF was tested in the test rack (Fig 3.2) at 160°C with gas feed rates of $\lambda=1.35$ for H_2 and $\lambda=2.5$ for air. The cell construction approach was the very same as reported and discussed in chapter 3 of this work. After reaching a cell temperature of 160°C, the single cell with Celtec P® 2000 MEA was operated under no load conditions (0 amps of current being drawn from the load connected to it). At the start, the OCV was observed to be close to 0.99 Volts and after 120 hours of no load test, the OCV came down to 0.90 Volts, as depicted in Fig 5.7. The performance of the cell was tested after this no load (or OCV) test, to gauge the degradation of the cell voltage caused due to OCV operation and was compared to its performance at start (or that before the said OCV test) (see Fig 5.9).

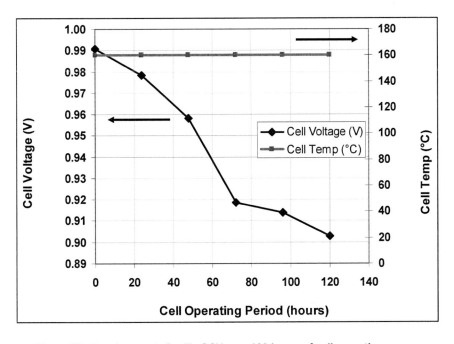

Figure 5.7: Development of cell's OCV over 120 hours of cell operation

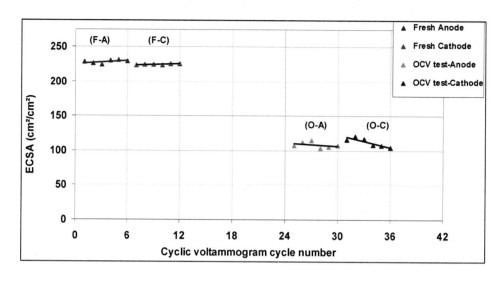

Figure 5.8: ECSA of a HTPEMFC before and after 120 hour OCV test. F-A: Fresh anode; F-C: Fresh cathode; O-A: OCV test-anode; O-C: OCV test-cathode.

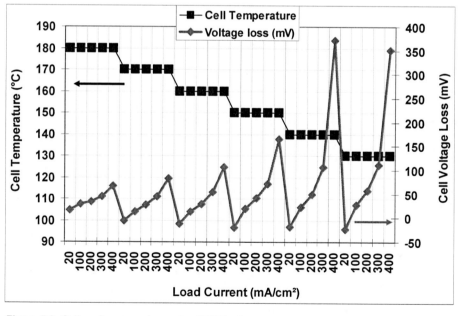

Figure 5.9: Cell performance loss after OCV test.

Figure 5.9 depicts the single cell HT PEMFC's voltage loss in mV at different temperatures after the said OCV test, at load currents of 20 mA/cm² - 400 mA/cm². In Figure 5.9, the X-axis shows the load current drawn from the single cell, the primary Y-axis shows the cell temperature and the secondary Y-axis shows the cell voltage loss, the reference being the cell performance before the OCV test at the same current and temperature. It can be clearly noticed from Fig 5.9 that the cell's performance loss increases with an increase in load current after the OCV test and further, this phenomenon is more pronounced at lower cell temperatures (130°C, 140°C and 150°C). That is, after the OCV test, operating the said HT- PEMFC at lower temperatures showed very low performance compared to its initial performance (before the OCV test at the same temperature and current). For instance, at 130°C and 140°C, after the OCV test, the cell had lost > 350 mV at 400 mA/cm². However, in the 150°C – 180°C temperature window, the cell performance loss was around 100 mV, at load currents of 20 mA/cm² - 300 mA/cm². This performance loss, which is a direct result of OCV test, suggests catalyst and catalyst support degradation, subsequently the loss of ECSA and possibly increased hydrogen crossover. CV scans were performed on the anode and cathode sides of the MEA, in the very same manner explained in chapter 4. The results are depicted in Fig 5.8. As can be seen from the figure, after the 120 hour OCV test, the ECSA on the anode and cathode sides were reduced to around 105 cm²/cm², from their initial value of around 225 cm²/cm². It can be concluded that OCV operation is detrimental to HT PEMFCs and must be avoided to ensure increased durability.

The cell performance loss was around 350 mV at 140°C (400 mA/cm² load current) as the ECSA was reduced to half its initial value after the OCV test. Therefore, the challenges involved with catalyst and catalyst supports in HT PEMFC MEAs needs to be addressed adequately. However, some groups are exploring other alternatives such as catalyst supports based on Tungsten Carbide [195,196] and graphene based clusters[197].

———

6 Conclusions and Outlook

6.1. Maximum theoretical thermodynamic cell voltage ($E_{th,T2}$) governing PEMFCs operated in the 25°C – 230°C was discussed, after deriving those values from the standard E_{th} values at STP (T_1) (Standard Temperature and Pressure of 25°C and 1 atm), fed with H_2 and O_2 as fuel and oxidant respectively. Similarly, the maximum theoretical thermodynamic cell efficiencies ($\eta_{th,T2}$) were derived from the calculated values of entropy and enthalpies at T_2 (30°C – 230°C). From these derivations, it can be said that a voltage loss of about 0.845 mV to 0.815 mV per 1° rise in operating temperature (in the temperature range 25°C to 90°C) is experienced by a PEMFC. Similarly, about 0.172 mV to 0.254 mV per 1° rise in cell temperature (in the temperature range 100°C to 230°C) is experienced by a PEMFC. This implies for instance, for a Low temperature (LT) PEMFC operating at 50°C, the E_{th} is 1.207 V, where as for a high temperature (HT) PEMFC operating at 160°C, the same is 1.152 V (a difference of 55 mV). Also, lower E_{th} values imply lower η_{max} **(maximum theoretical fuel cell efficiency)** under same operating conditions. From the calculated values of η_{max}, it was shown that about 0.053% to 0.057% loss from its initial efficiency of 82.9% (at STP) is experienced by a LT PEMFC for every 1° rise in cell temperature (in the 25°C to 90°C cell temperature range) and similarly, about 0.0161% to 0.0171% loss per every 1° rise in cell temperature (in the 100°C to 230°C cell temperature range) is experienced by the HT PEMFC. That implies that the *rate of fall* in theoretical cell efficiency is higher for LT PEMFCs when the cell temperature is raised from 25°C to 100°C, where as in HT PEMFCs (100°C to 230°C), the *rate of fall* (of theoretical cell efficiency) is lower. For instance, it was shown that for a PEMFC operating at 25°C, the maximum thermodynamic efficiency was 82.9%, where as it is 81.5% at 50°C and it is 79.9% at 80°C. At 160°C, the efficiency is 77.8% and at 180°C, the same is 77.5%. Further, when air is used in stead of pure oxygen in the fuel cell reaction considered here, a loss of 13.5 mV at 130°C and a loss of 15.2 mV at 180°C can be expected as a result of reduced O_2 partial pressure. The absolute pressure of the reactant gases fed into the HT PEMFC single cell studied in this work were 1.002 atm and 1.004 atm for H_2 and air respectively. In theory, according to the Nernst equation for cell potential, this would mean a rise in cell voltage of 17.1 µV (130°C) to 19.2 µV (180°C) due to the slight increase in H_2 partial pressure and a rise of 68 – 76 µV (130°C – 180°C) due to the 4 mbar rise in cathode side air partial pressure, which is negligible. However, the OCV (open circuit potential or cell voltage on no load) measured at cell voltage terminals is usually lower than these theoretical values.

6.2. Measured cell's OCV: As operating HTPEMFCs on OCV for prolonged periods of time is detrimental to its electrochemically active surface area (ECSA), the cell was not operated on OCV for more than 60 seconds in the single cell tests reported. However, the observed OCV in the 90°C – 170°C temperature range was 0.955 V - 0.985 Volts. These values are appropriate for relatively new MEAs (at beginning of life or BOL), after which the observed OCV was mostly around 0.8 V. The difference between the theoretical thermodynamic potential discussed earlier and the OCV measured at the cell terminals can be attributed to (a) loss caused due to lower O_2 partial pressure, (b) loss caused due to product water partial pressure, based on the Nernst relation for fuel cell reaction, (c) loss caused due to leakage current of the membrane considered at their respective operating temperatures and (d) loss caused due to phosphate anionic adsorption on catalyst sites. The E_{th} values at 130°C and 180°C are 1.1599 Volts and 1.1476 Volts respectively. When fed with H_2 and air with their absolute pressures of 1.002 atm and 1.004 atm respectively (as in the current case), the HT PEMFC's OCV (H_2/Air operation) should in theory be 1.146 Volts at 130°C and 1.132 Volts at 180°C. Whereas the measured OCV values were about 160 mV to 170 mV lower than those theoretical values. The loss due to product water partial pressure is negligible. For instance, the amount of water produced in a PEMFC single cell is around (0.33668x) ml/hr, where "x" is the load current drawn. This would mean about 93.5 nanolitres/second of water production due to fuel cell reaction. The cell voltage loss due to leakage currents is briefly discussed later, whereas the cell voltage loss due to phosphate anionic adsorption is a well known issue pertinent to pt-based electrodes exposed to H_3PO_4 electrolyte environments (from PAFC experiences) and is beyond the scope of this work. However, it can be stated that phosphate anions such as ($H_2PO_4^-$) and (HPO_4^{2-}) adsorb on pt-catalyst sites leading to cell voltage loss, which was believed to be structure (pt surface area) sensitive.

6.3. Fuel crossover overvoltage: Leakage currents (caused due to hydrogen permeation from anode to cathode side and oxygen permeation from cathode to anode side) in Celtec P® based membranes are in the 2-5 mA/cm² range. This implies a total leakage current of 200 - 500 mA for the cell studied here. Also for the cell with 1.88 mΩ of HF (high frequency) Impedance, this would mean a loss of 0.377 mV to 0.94 mV. However, at higher stoichiometries used for the single cell studied in the current work (λ=1.35 and λ=2.5 for H_2 and air respectively), the equivalent leakage currents of the membrane are assumed to be negligible.

6.4. Breakup of performance limiting losses: As load current is drawn from the HT PEMFC single cell, the cell voltage falls commensurate with the current drawn. This cell voltage loss at different loads can broadly be divided into 4 types of losses namely 1) Ohmic, 2) Concentration, 3) Activation and 4) Fuel crossover overvoltages. The fuel crossover loss was assumed to be negligible in the current work, as discussed earlier. The actual cell voltage at any load can be ascertained by deducting the first 3 types of losses from the cell's OCV at respective loads. Experiments were performed to determine these cell performance limiting losses systematically. Furthermore, some additional tests were performed to understand mainly the HT membranes and their behaviour under conditions of cell compression, hydration due to the influence of product water and are enumerated here:

6.4.1. HT PEMFC's ohmic overvoltages: Employing EIS (electrochemical impedance spectroscopy) analyser from Zahnermesstechnik, a) Cell's (cell assembly + MEA + cable) high frequency resistance (HFR) versus cell compression at RTP (room temperature and pressure), b) Cell's (cell assembly without MEA + 1GDL+ cable) HFR at different temperatures, c) Cell's (cell assembly + MEA + cable) HFR versus temperature, d) HFR of cell components from a 4-pole conductivity measurement device (a separate test) were measured in the laboratory. From these measurements, e) the cell's HFR with all the components except only membrane at different temperatures (cell assembly without MEA+2GDL+2CL+cable) was deduced and finally, the f) Membrane's HFR as seen by the cell at different temperatures was ascertained from the relation: c) - e). The measurements are summarized: a) Cell assembly's HFR versus cell compression from test (a): The HFR of the cell assembly was initially 37 mΩ at a cell compression of 0.75 N/mm². Then, the cell's compression was raised systematically. Until 14 bar, there no substantial fall in the cell's HFR, whereas at around 15 bar, the cell's HFR fell to 16.9 mΩ very rapidly. One dominant reason for this behaviour was that the contact resistances of the cell, namely, the BPHP/GDL, BPHP/CC, GDL/CL interfacial resistances would come down as the cell compression increases. The dimensional change of the membrane (at increased compression) and the subsequent gain in ionic conductivity (due to lower ionic resistance) is not a dominant phenomenon. Furthermore, from test (b) the cell's compression was fixed at 2 N/mm² and its HFR at various temperatures was measured by raising the cell assembly's temperature gradually from room temperature. The change in cell's HFR from 40°C to 160°C was 1.5 mΩ, which can be attributed mainly to the changes in electrical conductivity of the cell components such as graphite compound based bipolar half plates (BPHPs), gold plated current collector plates (CC). From test (d), using the 4-pole conductivity measurements, the HFR of the BPHPs(4 no.s)used was measured to be 0.844 mΩ, the HFR of the current collect plates(2 no.s) was measured to be 0.030 mΩ, th GDEs'(2 no.s) HFR was 0.0013 mΩ. This implies a total cell's HFR of 2.75 mΩ at 160°C or 137 mΩ-cm² of area specific cell resistance for a 50 cm² cell used here. Finally, from test (f), the membrane resistances were 2.02 mΩ at 140°C, 1.93 mΩ at 150°C and 1.88 mΩ at 160°C. The Celtec P® 2000 based membrane's HFR was expressed in the arrhenius form where the pre-exponential factor (σ_0) was found to be 13466 and the activation energy (Ea) was found to be 18484 J/mol. Based on these values of cell's HFR, the HT PEM single cell's voltage loss due to its ohmic (ionic and electronic) losses was determined at different load currents. For instance, at 130°C, on 400 mA/cm² load, the ohmic overvoltage was 58.4 mV, whereas at 180°C, the same was 53 mV. The lowest was 2.65 mV at 180°C and at 20 mA/cm². The highest was 58.4 mV at 130°C and 400 mA/cm².

6.4.1.1. Influence of membrane hydration: A HT PEMFC single cell with a i) highly doped membrane containing MEA and a ii) lowly doped membrane based MEA was tested in the 0 - 400 mA/cm² range. In the case of highly doped membrane containing cell, the fall in membrane's HFR was from 3.46 mΩ (40 mA/cm²) to 2.91 mΩ (400 mA/cm²) in case of H_2/air. With H_2/O_2 operation, it was 3.46 mΩ (40 mA/cm²) and 2.83 mΩ (400 mA/cm²). This slight decrease in the cell's HFR can be attributed to the improvement in proton conducting ability of the electrolyte under conditions of cell hydration, aided by water produced in the fuel cell reaction, the cell temperature being maintained constant in this case at 170°C. In the case of the lowly doped membrane based cell, the HFR was higher as can be expected. However, its HFR fell from 17.2 mΩ (0 mA/cm²) to 11.1 mΩ (20 mA/cm²) and further to 6.2 mΩ (400 mA/cm²) at the cell's operating temperature of 180°C. The improvements in membrane conductivity were much more pronounced at higher load currents (higher amounts of water produced). Although this phenomenon was not observed in the case of highly doped membrane containing MEA. In other words, it can be said that **water does play** a major role in these lowly doped (H_3PO_4/PBI doping < 10 moles/ mole of PBI repeat unit) membranes when compared to the highly doped (H_3PO_4/PBI doping > 20 moles/mole of PBI repeat unit) ones, although they are essentially anhydrous (proton conductivity not critically dependant on water). However, the absolute value of membrane conductivity for lowly doped membranes is about 2.5 times lower compared to the highly doped ones.

6.4.2. HT PEMFC's fuel transport overvoltages: By keeping the feed gas stoichiometries of 1.35 and 2.5 for H_2 and air respectively, load current was drawn gradually from the HT PEMFC single cell. The cell's limiting current was obtained by noting down the maximum current that can be drawn at these fuel/oxidant feed rates, when the cell voltage fell close to 0 volts. This maximum current, which is also called the cell's limiting current (i_L) was found to be 1600 mA/cm² at 180°C and it was 1480 mA/cm² at 130°C. Based on these (i_L) values, the mass transport overvoltages at different tempertures and load currents was determined. For instance, at 130°C, the mass transport overvoltage was 10.9 mV (400 mA/cm² load) and it was 11.2 mV at 180°C on the same load. These overvoltges are lower at lower loads. For instance, on 20 mA/cm², the mass transport overvoltages are 0.47 mV to 0.49 mV in the 130°C - 180°C temperature range.

6.4.3. HT PEMFC's Voltage at different load currents and temperatures: Single cell's voltage at load currents of (20 - 400 mA/cm²) and temperatures (130°C - 180°C) was studied by operating the cell in the test stand with gas feeds of λ=1.35 for H_2 and λ=2.5 for air, systematically. Operating the cell at lower temperatures (~ 130°C) might be interesting during the start up period, as these HT PEMFCs need to be first heated up from room temperature to the required cell temperature (typically 160°C). Operating the cell at 130°C is not attractive from the point of view of cell performance and tolerance to fuel impurities (such as CO in reformates). Whereas operating the cell close to 180°C may not be the best from the point of view of electrolyte retention, catalyst and catalyst support (carbon) stability. Thus, an operating cell temperature of 160°C might be a viable compromise between cell performance and tolerance to impurities, but offering higher durability. Therefore, taking the HT PEMFC operation at 160°C (with λ=1.35 for H_2 and λ=2.5 for air) as a reference, the rise in cell performance (expressed as rise in cell voltage per 1° rise in cell temperature) when the load current is raised from 5 amp (or 100 mA/cm²) to 20 amps (400 mA/cm²) in four steps was discussed. For instance, it was shown that, going from 160°C to 170°C (ΔT= 10 K), at a load current of 100 mA/cm², the performance **gain rate** was 0.69 mV/°C or a total performance **gain** of 6.9 mV. Whereas, at 400 mA/cm², under same conditions (160°C to 170°C), the performance gain rate was 0.81 mV/°C. Also, while going from 160°C to 180°C (ΔT= 20 K), the performance gain rate was 1.45 mV/°C (100 mA/cm²) and it was 1.03 mV/°C (400 mA/cm²), which translates to a total cell performance gain of 29 mV (1.45 * 20) in the first case and a gain of 20.6 mV (1.03 * 20) in the second case. Similarly, while going from 160°C down to 130°C (ΔT= -30 K), it was observed that at 100 mA/cm², the performance loss rate was 0.57 mV/°C, and at 400 mA/cm², it was 1.35 mV/°C. That implies a total performance loss of 17.19 mV (0.57 * 30) at 100 mA/cm² and a total loss of 40.59 mV (1.35 * 30) at 400 mA/cm² at cell's operating temperature of 130°C.

6.4.4. HT PEMFC's Activation overvoltages: The cell performance loss caused due to slow reaction kinetics at the triple-phase boundary (TPB) is also called the activation overpotential. These activation overpotentials (for anode and cathode reactions) for a HT PEMFC single cell operating at different temperatures were determined: Firstly, the measured cell voltages at different load currents and temperatures were corrected for ohmic overvoltages as well as the mass transport overvoltages discussed earlier and these corrected cell voltages were deducted from the cell's OCV. It was shown that the activation overpotentials are the highest losses followed by ohmic losses and fuel transport losses in their respective order. It was also shown that the same trend holds good for cell operation at temperatures of 130°C to 180°C, only that the activation overpotentials at lower temperatures are higher. For instance, at 180°C on 400 mA/cm² load, the single cell voltage was 0.6806 Volts and the activation overpotential was 0.2893 Volts. Whereas at the same load current, the cell voltage was 0.6244 Volts at 130°C and the activation overpotential was 0.3455 Volts. This difference of 56.2 mV can be mainly attributed to sluggish kinetics when the cell temperature is lowered to 130°C from 180°C, at 400 mA/cm².

6.5. Determination of kinetic parameters: The fuel cell's activation overvoltage is a function of variables such as exchange current density (i_0), the electron transfer co-efficient (α) and operating temperature (T). Tafel slopes were deduced from the cell performance curves (cell's activation overvoltages on x-axis versus logarithmic of load currents on y-axis). Subsequently, the (i_0) values were determined from the y-axis intercept of Tafel slopes at zero activation overpotentials (at x = 0). The Tafel slope was 106 mV/decade at 180°C and was 79 mV/decade at 130°C, which translates to a (i_0) value of 0.83 mA/cm² at 180°C and 0.04 mA/cm² at 130°C. The Tafel slopes and subsequent (i_0) values at other intermediate temperatures such as 140°C-170°C were in their respective range, where the Tafel slope and (i_0) at 180°C was the highest and the Tafel slope and the subsequent (i_0) at 130°C was the lowest in the measured temperature range. The transfer co-efficient (α) values were determined from Tafel slopes ($b = RT/\alpha nF$) at various temperatures. The (α) value was 0.42 at 180°C and was 0.50 at 130°C. The (α) values at other intermediate temperatures lies in the range between those at 180°C (min) and 130°C (max). Furthermore, the apparent exchange current densities (AECD- i_0) were determined from the nyquist plots of the cell impedance (EIS spectra) obtained at very low currents densities corresponding to activation overvoltages (η_{act}) < 10 mV, from the charge transfer resistance values at such low (η_{act}) values. The measured AECD values were 0.56 mA/cm² at 160°C and 1.91 mA/cm² at 180°C respectively. These AECDs translate to the activation overvoltages of 122 mV at 160°C (400 mA/cm² load), 102 mV at 170°C and 104 mV at 180°C on the same load. These AECD based activation overvoltages are about 2.5 times lower compared to Tafel slope based activation overvoltages discussed above. As the AECD values hold good only for cell operation close to OCV, these differences are expected.

6.6. CO tolerance of HT PEMFCs: HT PEMFC's tolerance to CO (carbon monoxide) of 1% to 20% in the anode stream was studied by means of single cell operation in the test stand. It was observed that at 180°C, a HT PEMFC fit with a Celtec P® 2000 MEA could offer a CO tolerance of up to 20% with a loss of about 150 mV (at 200 mA/cm² of load), whereas the same could not even stand 0.5% CO at 130°C. The cell had lost almost the entire cell voltage when fed with 1% CO into its anode stream of H_2 at 130°C. This implies, while designing HT PEMFC-liquid or gaseous fuel reformer coupled systems, care must be taken to avoid reformate feeds at cell temperatures below 130°C (with CO concentration > 0.1%). Furthermore, HT PEMFC single cell's performance was studied when fed with 1% CO with its anode stream of H_2 in the 130°C – 180°C temperature range and in the 20 mA/cm² - 400 mA/cm² load range, as it is more interesting to the reformer-coupled-HT PEMFC system configurations. It was shown that at temperatures of more than 170°C, 1% CO in the anode stream did not result in a significant cell voltage loss, whereas at temperatures below 150°C, the voltage loss was considerable. For instance, at 180°C, with 1% CO fed into the anode stream of H_2 (λ = 1.35), cathode being fed with air (λ = 2.5), while going from 20 mA/cm² to 400 mA/cm² load, the cell had lost 0.6 mV to 4.4 mV respectively (compared to the case of pure H_2 feed at the same (λ of 1.35), which is negligible. But at 150°C, with 1% CO feed into the anode stream of H_2 (λ = 1.35), air feed rate being the same (as at 180°C), while going from 20 mA/cm² to 400 mA/cm², the cell voltage loss was 5.5 mV to 52.2 mV respectively, compared to the case of pure H_2 feed, which is considerable. Furthermore, at 130°C, with 1% CO feed into the anode stream, air feed rate being the same as in the previous two cases, while going from 20 mA/cm² to 400 mA/cm² load, the cell had lost 21.1 mV to 534.1 mV respectively, compared to the case of pure H_2 feed which is significantly high. The β values were calculated from the relation $(\Delta V)/(RT/nF)$, where ΔV stands for cell voltage loss when fed with 1% CO compared to 100% H_2 feed, R, and F are standard constants, n = 2 and T is the cell's operating temperature in Kelvin. These β values were found to vary between 0.031 (at 20 mA/cm²) at 180°C to 0.225 (at 400 mA/cm²) at 180°C. At 160°C, β values were 0.096 (20 mA/cm²) and 0.954 (400 mA/cm²). Whereas at 150°C, β values were 0.302 (20 mA/cm²), 0.603 (100 mA/cm²), 1.053 (200 mA/cm²), 1.975 (300 mA/cm²) and 2.863 (400 mA/cm²). That implies, operating the cell when fed with 1% CO into its anode stream at β values of less than 1 is recommended to ensure cell operation without significant performance loss.

6.7. HT PEMFC stack's performance when fed with reformates: A short HT PEMFC stack (consisting of 12 individual cells with Celtec P® 1000 MEAs), with a total active area of about 600 cm² was studied. This was an air cooled short stack with a nomial power output of around 165 W_{el} at about 0.6 Volts / cell. Every alternate cell had cooling channels integrated into the graphite compound based BPHPs used. The 12-Cell stack was then tested systematically with H_2 and air first and subsequently with three types of synthetic reformates namely a) Methanol reformate (H_2:75%; CO:1%; CO_2: 24%), b) Propane steam reformate (H_2:54.8%; CO:0.55%; CH_4:1.38%; CO_2:16.47%; H_2O: 26.78%) and c) Natural gas autothermal reformate (H_2:35%; CO:0.2%; CO_2:12%; N_2:52.8%) at stack temperature of 165°C, with feed gas stoichiometries of 1.2 and 2.0 respectively for fuel and air. The 12 cell stack delivered 78.4 Watts of electrical power at 200 mA/cm² of load current, and 112.4 Watts at 300 mA/cm² and about 166 Watts at 500 mA/cm², when operated at 165°C and fed with H_2 and air at mentioned gas feed rates. But when fed with synthetic reformates, the power loss was more pronounced in case of natural gas autothermal reformate, when compared to the methanol steam reformate (which is commensurate with the H_2 content of the reformate). At higher load currents, this phenomenon was even more pronounced. For instance, the power loss with natural gas reformate at 200 mA/cm² load was 3.6 Watts, whereas the power loss with the same reformate at 500 mA/cm² was 20 Watts (pure H_2 and air feed at λ = 1.2 and 2.0 respectively being the reference case). Also, up to about 370 mA/cm² or until the voltage level per cell was around 0.6 V, the stack performance was not significantly affected by the reformate constituents (also because the CO concentration was below 1% in the reformate). The HT PEMFC stack exhaust gases from its anode were analysed using a gas analyser from Rosemount®, Germany to check for the concentration of constituents such as CO, CO_2 and H_2. From that study, it was observed that the CO, which was present in the synthetic reformates being fed into the HT PEMFC stack had exited in its entirety from the anodic outlet. In conclusion, it can be said that, with the reformate feeds consisting more of inert gas composition (for instance N_2 and less of H_2), the dilution effect (lower H_2 partial pressure) was clearly seen from the lower stack performance.

6.8. Electrolyte (PA) loss : The long term performance of two HT PEM single cells containing Celtec P® 1000 MEAs operated at 170°C and 160°C respectively was studied for 1000 hours, to check for electrolyte loss patterns. It was demonstrated that operating a HT PEMFC on H_2 and air at 170°C showed higher initial performance compared to the one operating at 160°C under same gas feed and load current conditions, but after 900 hours of cell operation, the latter outperformed the former. For instance, after 900 hours, the cell (that was operated at 170°C) had lost 40 mV compared to its voltage at BOL, whereas the cell that was operated at 160°C had lost 29 mV from its voltage at BOL. That is, the cell's performance loss rate was 44 μV/hr in the first case, whereas the same was 32 μV/hr in the second case. It can be concluded that the electrolyte loss in HT PEMFCs based on PBI/ H_3PO_4 systems can be higher at higher cell operating temperatures. A 10°K rise in temperature (from 160°C or 433.15°K) resulted in an electrolyte loss rate of > 2.0 times its value at 160°C. The electrolyte loss rate was around 0.45 μg/m²/s in the case of 170°C cell, whereas its was around 0.20 μg/m²/s in the case of 160°C cell. However, it must be noted that the CO (Carbon Monoxide) tolerance of a HT PEMFC operating at 170°C will be higher compared to the cell operating at 160°C. For instance, a single cell operating at 170°C could tolerate 1% CO present in the fuel gas, with a loss of about 3.2 mV in cell voltage at 200 mA/cm² (compared to the case of 100% hydrogen feed). Whereas the same at 160°C could lose about 12.1 mV, under same conditions, when Celtec ® P 1000 MEAs are used in a HT PEMFC, an enumerated earlier.

6.9. 2100 hour test performance test: The long term performance of a HT PEMFC single cell was studied for about 2100 hours which was consisted of a commercially available Celtec P® 2000 MEA, commercially available bipolar half plates (BPHPs), gaskets, etc. It was observed that the various operating regimes such as constant load operation, temperature and load cycling, cell operation on no load (higher cell voltage) and (HT PEMFC short-circuited with a resistor) had their respective influence on cell performance degradation. However, the dominant degradation mode was the OCV operation. Although Celtec P® 2000 based MEAs are expected to be more stable compared to Celtec P® 1000 based ones, the observed performance loss was 10 µV/hr on constant load and an average of 20 µV/hr (with constant load, load cycling and temperature cycling, shutdown and start-up phases included) for the said 2100 hours of cell operation. A loss of 100 mV implies 5000 hours of cell operation at the mentioned performance loss rate of 20 µV/hr. At a performance loss rate of 10 µV/hr, the cell could operate for 10,000 hours before it loses 100 mV. In a scenario which includes load and thermal cycles along with other start up and shut down procedures, this implies anywhere between 5,000 and 10,000 hours of HT PEM fuel cell operation before the same could lose 10% of its initial performance. Also, from the cell performance (I-V curves) studied before and after the 2100 hour test, it can be said that the cell had lost about 2.3 mV/amp from its initial cell potential at BOL in the 1-20 ampere current range (for e.g., a total of 23 mV at 10 amps or 46 mV at 20 amps) at EOL. The total cell voltage loss during those 2100 hours was 42.2 mV (0.6610 – 0.6188 V) on an average load of 200 mA/cm², at an average cell temperature of 160°C.

6.10. Determination of ECSA (Electrochemically active surface area): The electrochemically active surface area of the Celtec P® 2000 MEA before and after the 2000 hour test described above is analysed using cyclic voltammetry (CV) technique. Both anode and cathode electrodes were studied employing the procedure described here: The electrode of interest or the working electrode (WE) was fed with 100 ml/min of nitrogen and the counter electrode (CE) was fed with 100 ml/min of hydrogen. Zahnermesstechnik's IM6 was connected to the fuel cell, in a two-electrode configuration, where the WE was connected to IM6's positive terminal and the CE to the negative terminal. Firstly, the cathode of fresh Celtec P 2000 MEA was considered as WE and 6 CV scans were performed with a sweep rate of 50 mV/s in the potential range 50 mV to 1000 mV versus the CE. From these 6 CV scans, the charge density q_{pt} due to H-adsorption during the reverse scan was determined, following the relation ($q_{pt} = H/vA$), where q_{pt} is charge density (in coloumbs/cm²), H is area under H-adsorption curve from CV scans (volt.amps), v is sweep rate in volts/second, A is catalyst layer's geometrical area. The non-faradaic part that arises due to charging and discharging of electric double layer capacitance of electrode/electrolyte interface, was excluded from q_{pt}, as this part contributes to charge accumulation but not to charge transfer. Also, after the 2000 hour test,

CV scans were performed on the cathode as well as on the anode side of the used MEA. The determined charge densities were normalized to (Γ = 210 µC/cm²-pt), where Γ stands for charge required to reduce a monolayer of protons on Pt using the relation [$ECSA_{pt} = q_{pt}/ \Gamma.L$], where, L is the pt loading of the electrode in g_{pt}/cm² (0.004 g/cm² in this case). From these tests, it was evident that there was no significant change in the anode side ECSA after the 2000 hour test, whereas on the cathode side, the ECSA before the long term test was 225 cm²/cm² and after the 2000 hour test, it was reduced to 180 cm²/cm², which is a 20% reduction in ECSA. It may be recalled that the cell had lost 42.2 mV during those 2100 hours, which is a 6.5% reduction in cell power.

6.11. Electrolyte loss during the long term test: During the 2100 hour operation of a single cell HT PEMFC equipped with Celtec P 2000 MEA from BASF, during load and temperature cycling phases, about 3.48 mg of phosphoric acid per litre of water was collected which corresponds to 0.65 µg/m²/s of PA loss (or electrolyte loss in µg per second of HT PEMFC operation per square meter of the membrane). During constant load operation, a loss of 2.27 mg of phosphoric acid per litre of water was collected, which corresponds to the electrolyte loss of 0.42 µg/m²/s. An amount of 0.5 µg/m²/s implies that the corresponding membrane could operate for about 40,000 hours before it could reach the end of life, which could be defined as the time taken to lose 10% of its initial performance (cell voltage) measured at the beginning of life. In all the phases of HT PEMFC operation, the PA loss should not exceed 0.5 µg/m²/s to avoid unacceptable cell performance loss caused due to electrolyte loss. This can be achieved by closing the in and outlets of the HT PEMFC during heating up phase and intermittent shutting off of the in and outlets during cooling down phase. While a HT PEMFC is cooled down from its 170°C to 100°C or 40°C, water condensation takes place inside the connected pipes which must be flushed out, to avoid electrolyte loss. Also, the outlets of both anode and cathode of a HT PEMFC could be directed through a container filled partially with ceramic pellets (or a condenser with a diaphragm based separator) by means of which, the liquid water collected in the bottle will not be able to re-enter the cell during the start-up phase.

6.12. Degradation issues: A careful examination (using SEM or scanning electron microscpy at the university of Duisburg-Essen) of the surface of the bipolar half plates at various points across their entire area has revealed that these plates were stable after the long term operation. However, deposits of PA (H_3PO_4 leached from the cell's membrane) were observed on their surface.. It is likely that the excess acid which could have been squeezed out of the highly doped MEA, at the beginning of life (BOL) could have deposited itself on the surface of the bipolar half plate. Further, the electrolyte (PA) that was leached out of the HT

MEA, at the rate of about 0.2 μg/m²s could have formed a thin phosphate layer on the surface of the BPHPs.

6.13. OCV (No load operation of HT PEMFC single cell) test: A single cell with a Celtec P 2000 MEA from BASF was tested in a test rack at 160°C with H_2 and air. The cell construction approach was the very same as reported and discussed in chapter 3 of this work. After reaching a cell temperature of 160°C, this single cell was operated under no load conditions (0 amps of current being drawn from the load connected to it). At the start, the OCV was observed to higher than 0.97 Volts and after 120 hours of no load test, the OCV came down to 0.90 Volts. The performance (voltage) of the cell was tested (at different loads) after this no load (or OCV) test, to gauge the degradation of the cell caused due to OCV operation and was compared to its initial performance. The single cell HT PEMFC performance loss in mV at different temperatures after the said OCV test, at load currents of 20 mA/cm² - 400 mA/cm² was recorded. For instance, at 130°C and 140°C, after the OCV test, the cell had lost > 350 mV at 400 mA/cm². However, in the 150°C – 180°C temperature window, the cell performance loss was around 100 mV, at load currents of 20 mA/cm² - 300 mA/cm². This performance loss, which is a direct result of OCV test, suggests catalyst and catalyst support degradation and subsequently the loss of ECSA and possibly increased hydrogen crossover. CV scans were performed on the anode and cathode sides of the MEA, in the very same manner explained in the earlier part. It was observed that after the 120 hour OCV test, the ECSA on the anode and cathode sides were reduced to around 105 cm²/cm², from their initial value of around 225 cm²/cm². It can be concluded that OCV operation is detrimental to HT PEMFCs and must be avoided to ensure increased durability. The commercially available MEAs to be used in HT PEMFCs were studied systematically. Their performance at different temperatures and loads was analysed. Although many degradation processes and phenomena were discussed with experiments in this work, it can be said that HT PEMFCs hold a great promise. Because, they are simple to operate, with a potential to offer a combined heat and power efficiency of > 70% in most cases. The issues related to catalyst and catalyst support durability still remain a challenge. The attempts by many research groups to realise non-Pt based catalysts (or catalysts with low Pt loading), might make these HT PEMFC related MEAs more attractive. In the future energy scenario of the 21st century, where decentralized energy patterns might be the order of the day, HT PEMFCs might find an important share in the total power production scheme. The current work aims to make explicit, the pros and cons of this promising technology, with a primary focus on the long term performance and durability of HT PEMFCs to be used as combined heat and power units.

7 List of Figures

8 List of Tables

Appendix I

Appendix I - Table 1

Standard Entropy Values at reference temperaure T_1			
at T_1 = 298.15°K			
$\Delta S°$ at T_1 (a,b,c,d) :		- [$\Delta G°$ a,b,c,d T_1 - $\Delta H°$ a,b,c,d T_1]/ T_1	
$\Delta S°$ at T_1 (a) :	H_2 (g)	0	J/mole.K
$\Delta S°$ at T_1 (b) :	O_2 (g)	0	J/mole.K
$\Delta S°$ at T_1 (c) :	H_2O (g)	-44.427	J/mole.K
$\Delta S°$ at T_1 (d) :	H_2O (Liquid)	-163.43	J/mole.K

Appendix I - Table 2

Standard Enthalpy Values at reference temperature T_1			
at T1 = 298.15°K			
$\Delta H°$ at T_1 (a,b,c,d) :			
$\Delta H°$ at T_1 (a) :	H_2 (g)	0,00	J/mole
$\Delta H°$ at T_1 (b) :	O_2 (g)	0,00	J/mole
$\Delta H°$ at T_1 (c) :	H_2O (g)	-241818.00	J/mole
$\Delta H°$ at T_1 (d) :	H_2O (Liquid)	-285830.00	J/mole

Appendix I - Table 3

Standard Enthalpy & Gibbs Free energy at reference temperature T_1			
Chemical Species	State	$\Delta H°$ at 25°C	$\Delta G°$ at 25°C
		J/mol	J/mol
H_2 (a)	Gas	0	0
O_2 (b)	Gas	0	0
H_2O (c)	Gas	-241818	-228572
H_2O (d)	Liquid	-285830	-237129

Appendix I - Table 4

Coefficients A,B,C and D from Thermodynamic Tables					
Chemical Species	State	A	10^3 B	10^6 C	10^{-5} D
H_2	Gas	3.249	0.422	0	0.083
O_2	Gas	3.639	0.506	0	-0.227
H_2O	Gas	3.47	1.45	0	0.121
H_2O	Liquid	8.71	1.25	-0.18	0.00

Appendix I - Table 5

Entropy calculations

$\Delta S°$ at T_2 (a) :	$[\Delta S°$ (a) at $T_1] + R^* [A \ln (T) + B T + C/2\ T^2 - D/2\ T^{-2}]$ where T_1: 298.15°K
Gas Constant	$R =$ 8.314 J/mol.K

Operating temperature		For Gas (a) or H_2 in gaseous form					
°C	°K	Part A	Part B	Part C	Part D	Entropy change	$\Delta S°$ at T_2 (a) (J/mol.K)
25	298.15						
30	303.15	0.05403417	0.00211	0	-0.001527301	0.479480641	0.479480641
35	308.15	0.10718439	0.00422	0	-0.002980861	0.950998942	0.950998942
40	313.15	0.15947909	0.00633	0	-0.004365352	1.414830337	1.414830337
45	318.15	0.2109454	0.00844	0	-0.005685083	1.871235989	1.871235989
50	323.15	0.26160914	0.01055	0	-0.006944029	2.320463766	2.320463766
55	328.15	0.31149497	0.01266	0	-0.008145867	2.762749145	2.762749145
60	333.15	0.36062641	0.01477	0	-0.009294	3.198316028	3.198316028
65	338.15	0.40902593	0.01688	0	-0.01039158	3.627377498	3.627377498
70	343.15	0.45671503	0.01899	0	-0.011441533	4.050136515	4.050136515
75	348.15	0.50371426	0.0211	0	-0.012446574	4.466786552	4.466786552
80	353.15	0.55004329	0.02321	0	-0.013409229	4.877512183	4.877512183
85	358.15	0.59572097	0.02532	0	-0.014331849	5.282489628	5.282489628
90	363.15	0.64076536	0.02743	0	-0.015216622	5.681887254	5.681887254
100	373.15	0.72902286	0.03165	0	-0.016880663	6.464580005	6.464580005
105	378.15	0.77226854	0.03376	0	-0.017663617	6.848176614	6.848176614
110	383.15	0.81494616	0.03587	0	-0.018416119	7.22679714	7.22679714
115	388.15	0.85707043	0.03798	0	-0.01913973	7.600577016	7.600577016
120	393.15	0.89865554	0.04009	0	-0.019835907	7.969646152	7.969646152
125	398.15	0.9397151	0.0422	0	-0.020506022	8.334129236	8.334129236
130	403.15	0.98026224	0.04431	0	-0.021151359	8.694146012	8.694146012
135	408.15	1.02030959	0.04642	0	-0.021773125	9.049811541	9.049811541
140	413.15	1.05986931	0.04853	0	-0.022372453	9.401236443	9.401236443
145	418.15	1.09895315	0.05064	0	-0.022950411	9.748527126	9.748527126
150	423.15	1.1375724	0.05275	0	-0.023508002	10.091786	10.091786
155	428.15	1.175738	0.05486	0	-0.024046173	10.43111167	10.43111167
160	433.15	1.21346047	0.05697	0	-0.024565815	10.76659915	10.76659915
165	438.15	1.25074999	0.05908	0	-0.025067768	11.09833998	11.09833998
170	443.15	1.28761638	0.06119	0	-0.025552828	11.42642246	11.42642246
175	448.15	1.32406914	0.0633	0	-0.026021742	11.75093176	11.75093176
180	453.15	1.36011744	0.06541	0	-0.026475221	12.07195009	12.07195009
185	458.15	1.39577016	0.06752	0	-0.026913933	12.38955682	12.38955682
190	463.15	1.43103589	0.06963	0	-0.027338514	12.70382863	12.70382863
195	468.15	1.46592294	0.07174	0	-0.027749564	13.01483959	13.01483959
200	473.15	1.50043936	0.07385	0	-0.028147651	13.32266134	13.32266134
205	478.15	1.53459294	0.07596	0	-0.028533316	13.62736314	13.62736314
210	483.15	1.56839123	0.07807	0	-0.028907069	13.92901202	13.92901202
215	488.15	1.60184154	0.08018	0	-0.029269396	14.22767284	14.22767284
220	493.15	1.63495097	0.08229	0	-0.029620758	14.52340838	14.52340838
225	498.15	1.66772639	0.0844	0	-0.029961594	14.81627948	14.81627948
230	503.15	1.70017448	0.08651	0	-0.030292319	15.10634507	15.10634507

Δ S° at T_2 (b) :		[ΔS° (b) at T_1] + R* [A ln (T) + B T + C/2 T² - D/2 T⁻²] where T_1: 298.15°K					
Gas Constant		R =	8.314	J/mol.K			
Operating temperature		For Gas (b) or O_2 in gaseous form					
°C	°K	Part A	Part B	Part C	Part D	Entropy change	Δ S° at T_2 (b) (J/mol.K)
25	298.15						
30	303.15	0.06052027	0.00253	0	0.004177075	0.489471747	0.489471747
35	308.15	0.12005047	0.00506	0	0.008152475	0.972388783	0.972388783
40	313.15	0.17862247	0.00759	0	0.011938975	1.448909871	1.448909871
45	318.15	0.23626664	0.01012	0	0.015548358	1.919189463	1.919189463
50	323.15	0.2930119	0.01265	0	0.0189915	2.38337772	2.38337772
55	328.15	0.34888587	0.01518	0	0.022278456	2.841620557	2.841620557
60	333.15	0.40391489	0.01771	0	0.02541853	3.294059703	3.294059703
65	338.15	0.45812415	0.02024	0	0.028420346	3.74083277	3.74083277
70	343.15	0.5115377	0.02277	0	0.031291903	4.182073339	4.182073339
75	348.15	0.56417857	0.0253	0	0.03404063	4.61791105	4.61791105
80	353.15	0.6160688	0.02783	0	0.036673434	5.048471692	5.048471692
85	358.15	0.66722949	0.03036	0	0.039196743	5.473877309	5.473877309
90	363.15	0.71768087	0.03289	0	0.041616544	5.894246299	5.894246299
100	373.15	0.81653253	0.03795	0	0.046167595	6.720330383	6.720330383
105	378.15	0.86496929	0.04048	0	0.048308928	7.12626498	7.12626498
110	383.15	0.91276979	0.04301	0	0.050366977	7.527602161	7.527602161
115	388.15	0.95995054	0.04554	0	0.052346007	7.924443653	7.924443653
120	393.15	1.0065274	0.04807	0	0.054250012	8.316888152	8.316888152
125	398.15	1.05251562	0.0506	0	0.056082736	8.705031429	8.705031429
130	403.15	1.09792992	0.05313	0	0.057847693	9.088966417	9.088966417
135	408.15	1.14278442	0.05566	0	0.059548184	9.468783316	9.468783316
140	413.15	1.18709277	0.05819	0	0.061187311	9.844569674	9.844569674
145	418.15	1.23086811	0.06072	0	0.062767991	10.21641049	10.21641049
150	423.15	1.27412311	0.06325	0	0.06429297	10.58438827	10.58438827
155	428.15	1.31686999	0.06578	0	0.065764835	10.94858316	10.94858316
160	433.15	1.35912055	0.06831	0	0.067186024	11.30907299	11.30907299
165	438.15	1.40088619	0.07084	0	0.068558836	11.66593335	11.66593335
170	443.15	1.4421779	0.07337	0	0.069885444	12.01923769	12.01923769
175	448.15	1.48300634	0.0759	0	0.071167897	12.36905738	12.36905738
180	453.15	1.52338176	0.07843	0	0.072408133	12.71546179	12.71546179
185	458.15	1.56331413	0.08096	0	0.073607986	13.05851832	13.05851832
190	463.15	1.60281305	0.08349	0	0.074769189	13.39829254	13.39829254
195	468.15	1.64188784	0.08602	0	0.075893385	13.73484818	13.73484818
200	473.15	1.6805475	0.08855	0	0.07698213	14.06824722	14.06824722
205	478.15	1.71880077	0.09108	0	0.078036899	14.39854997	14.39854997
210	483.15	1.7566561	0.09361	0	0.079059091	14.7258151	14.7258151
215	488.15	1.79412169	0.09614	0	0.080050035	15.05009968	15.05009968
220	493.15	1.83120547	0.09867	0	0.081010989	15.37145928	15.37145928
225	498.15	1.86791515	0.1012	0	0.081943154	15.68994799	15.68994799
230	503.15	1.90425821	0.10373	0	0.082847667	16.00561846	16.00561846

Δ S° at T₂ (c) :		[ΔS° (c) at T₁] + R* [A ln (T) + B T + C/2 T² - D/2 T²] where T₁: 298.15°K					
Gas Constant		R =	8.314	J/mol.K			
Operating temperature		For Gas (c) or H₂O in gaseous form					
°C	°K	Part A	Part B	Part C	Part D	Entropy change	Δ S° at T₂ (c) (J/mol.K)
25	298.15						
30	303.15	0.057709629	0.00725	0	-0.002226547	0.558585863	-43.86841414
35	308.15	0.114475168	0.0145	0	-0.004345592	1.108428797	-43.3185712
40	313.15	0.170327009	0.02175	0	-0.006363947	1.649838111	-42.77716189
45	318.15	0.225294102	0.029	0	-0.008287891	2.183106692	-42.24389331
50	323.15	0.27940404	0.03625	0	-0.010123223	2.708512158	-41.71848784
55	328.15	0.332683146	0.0435	0	-0.0118753	3.226317918	-41.20068208
60	333.15	0.385156548	0.05075	0	-0.013549084	3.73677413	-40.69022587
65	338.15	0.436848254	0.058	0	-0.015149171	4.240118591	-40.18688141
70	343.15	0.48778121	0.06525	0	-0.016679825	4.736577543	-39.69042246
75	348.15	0.537977369	0.0725	0	-0.018145006	5.226366425	-39.20063358
80	353.15	0.587457746	0.07975	0	-0.019548394	5.709690548	-38.71730945
85	358.15	0.636242466	0.087	0	-0.020893418	6.186745738	-38.24025426
90	363.15	0.68435082	0.09425	0	-0.022183268	6.657718909	-37.76928109
100	373.15	0.778611675	0.10875	0	-0.024609159	7.582125511	-36.84487449
105	378.15	0.824798967	0.116	0	-0.025750574	8.035892881	-36.39110712
110	383.15	0.870379551	0.12325	0	-0.026847596	8.484247001	-35.942753
115	388.15	0.915369161	0.1305	0	-0.027902497	8.927337568	-35.49966243
120	393.15	0.959782925	0.13775	0	-0.028917407	9.365308061	-35.06169194
125	398.15	1.003635398	0.145	0	-0.029894322	9.798296088	-34.62870391
130	403.15	1.04694059	0.15225	0	-0.030835114	10.2264337	-34.2005663
135	408.15	1.089711993	0.1595	0	-0.031741543	10.6498477	-33.7771523
140	413.15	1.131962606	0.16675	0	-0.032615263	11.0686599	-33.3583401
145	418.15	1.17370496	0.174	0	-0.033457828	11.48298742	-32.94401258
150	423.15	1.214951137	0.18125	0	-0.034270702	11.89294287	-32.53405713
155	428.15	1.255712794	0.1885	0	-0.035055264	12.29863464	-32.12836536
160	433.15	1.296001184	0.19575	0	-0.035812814	12.70016708	-31.72683292
165	438.15	1.335827169	0.203	0	-0.036544578	13.0976407	-31.3293593
170	443.15	1.375201244	0.21025	0	-0.037251712	13.49115238	-30.93584762
175	448.15	1.414133549	0.2175	0	-0.037935311	13.8807955	-30.5462045
180	453.15	1.452633889	0.22475	0	-0.038596406	14.26666017	-30.16033983
185	458.15	1.490711743	0.232	0	-0.039235975	14.64883332	-29.77816668
190	463.15	1.528376283	0.23925	0	-0.039854942	15.0273989	-29.3996011
195	468.15	1.565636385	0.2465	0	-0.040454183	15.40243799	-29.02456201
200	473.15	1.602500643	0.25375	0	-0.041034528	15.77402891	-28.65297109
205	478.15	1.63897738	0.261	0	-0.041596761	16.14224741	-28.28475259
210	483.15	1.675074658	0.26825	0	-0.04214163	16.50716672	-27.91983328
215	488.15	1.71080029	0.2755	0	-0.042669842	16.86885768	-27.55814232
220	493.15	1.746161851	0.28275	0	-0.043182069	17.22738885	-27.19961115
225	498.15	1.781166688	0.29	0	-0.04367895	17.58282663	-26.84417337
230	503.15	1.815821924	0.29725	0	-0.044161091	17.93523529	-26.49176471

ΔS° at T₂ (d) :	[ΔS° (c) at T₁] + R* [A ln (T) + B T + C/2 T² - D/2 T⁻²] where T₁: 298.15°K					
Gas Constant	R =		8.314	J/mol.K		
Operating temperature	For Liquid (d) or H₂O in liquid form					
°C	°K	Part A	Part B	Part C	Part D	Entropy change	ΔS° at T₂ (d) (J/mol.K)
25	298.15						
30	303.15	0.14488942	0.00625	-0.00027059	0	1.254323491	-162.1756765
35	308.15	0.28740855	0.0125	-0.00054567	0	2.488902964	-160.941097
40	313.15	0.42763369	0.01875	-0.00082526	0	3.704372826	-159.7256272
45	318.15	0.56563753	0.025	-0.00110934	0	4.901337337	-158.5286627
50	323.15	0.70148934	0.03125	-0.00139793	0	6.080372484	-157.3496275
55	328.15	0.83525521	0.0375	-0.00169101	0	7.242027724	-156.1879723
60	333.15	0.96699823	0.04375	-0.0019886	0	8.386827584	-155.0431724
65	338.15	1.09677867	0.05	-0.00229068	0	9.515273148	-153.9147269
70	343.15	1.22465415	0.05625	-0.00259727	0	10.62784344	-152.8021566
75	348.15	1.35067978	0.0625	-0.00290835	0	11.72499668	-151.7050033
80	353.15	1.47490832	0.06875	-0.00322394	0	12.8071715	-150.6228285
85	358.15	1.59739031	0.075	-0.00354402	0	13.87478802	-149.555212
90	363.15	1.71817416	0.08125	-0.00386861	0	14.9282489	-148.5017511
100	373.15	1.95483139	0.09375	-0.00453128	0	16.99423263	-146.4357674
105	378.15	2.0707921	0.1	-0.00486936	0	18.00748169	-145.4225183
110	383.15	2.18522958	0.10625	-0.00521195	0	19.00802914	-144.4219709
115	388.15	2.29818332	0.1125	-0.00555903	0	19.99620337	-143.4337966
120	393.15	2.40969131	0.11875	-0.00591062	0	20.97232019	-142.4576798
125	398.15	2.51979008	0.125	-0.0062667	0	21.9366834	-141.4933166
130	403.15	2.62851482	0.13125	-0.00662729	0	22.88958545	-140.5404145
135	408.15	2.73589939	0.1375	-0.00699237	0	23.83130796	-139.598692
140	413.15	2.84197643	0.14375	-0.00736196	0	24.76212228	-138.6678777
145	418.15	2.94677741	0.15	-0.00773604	0	25.68228994	-137.7477101
150	423.15	3.05033265	0.15625	-0.00811463	0	26.59206318	-136.8379368
155	428.15	3.15267143	0.1625	-0.00849771	0	27.49168531	-135.9383147
160	433.15	3.25382199	0.16875	-0.0088853	0	28.3813912	-135.0486088
165	438.15	3.35381161	0.175	-0.00927738	0	29.26140761	-134.1685924
170	443.15	3.45266664	0.18125	-0.00967397	0	30.13195359	-133.2980464
175	448.15	3.55041253	0.1875	-0.01007505	0	30.99324082	-132.4367592
180	453.15	3.6470739	0.19375	-0.01048064	0	31.84547392	-131.5845261
185	458.15	3.74267455	0.2	-0.01089072	0	32.6888508	-130.7411492
190	463.15	3.83723751	0.20625	-0.01130531	0	33.52356289	-129.9064371
195	468.15	3.93078507	0.2125	-0.01172439	0	34.34979548	-129.0802045
200	473.15	4.02333879	0.21875	-0.01214798	0	35.16772795	-128.2622721
205	478.15	4.11491958	0.225	-0.01257606	0	35.97753401	-127.452466
210	483.15	4.20554767	0.23125	-0.01300865	0	36.77938196	-126.650618
215	488.15	4.29524269	0.2375	-0.01344573	0	37.5734349	-125.8565651
220	493.15	4.38402364	0.24375	-0.01388732	0	38.35985095	-125.0701491
225	498.15	4.47190899	0.25	-0.0143334	0	39.13878342	-124.2912166
230	503.15	4.5589166	0.25625	-0.01478399	0	39.91038107	-123.5196189

Fuel cell reaction entropy	$\Delta S°_{rxn}$ at $T_2 = \Delta S°(H_2O)$ at $T_2 - \Delta S°(H_2)$ at $T_2 - 0.5*\Delta S°(O_2)$ at T_2

Fuel cell reaction entropy is derived from the calculated values shown in Table 5 to Table 8, using the relation shown above and is plotted below (Figures 1 and 2):

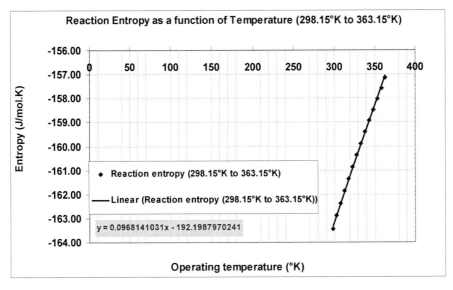

Appendix I – Figure 1

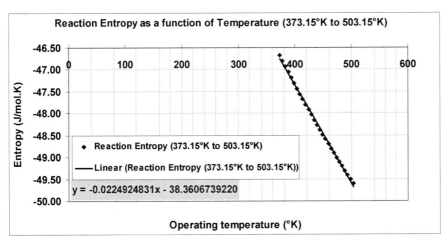

Appendix I – Figure 2

Conclusion: PEM fuel cell's reaction entropy (ΔS°_{rxn}) (J/mol.K)

At (25°C to 90°C or 298.15°K to 363.15°K) = 0.09681410313257030*T - 192.1987970240730

at (100°C to 230°C or 373.15°K to 503.15°K) = -0.02249248312077050*T – 38.36067392197640

(T = Temperature in Kelvin)

Appendix I - Table 9

Enthalpy calculations							
Δ H° (a) at T_2 :	**[Δ H° (at T_1=298.15°K)] + R*[A*(T_2-T_1) + B/2*(T_2^2-T_1^2)+C/3*(T_2^3-T_1^3) - D(1/T_2-1/T_1)]**						
Gas Constant	R =	8.314	J/mol.K				
Operating Temperature		**For Gas (a) or H_2 in gaseous form**					
°C	°K	Part A	Part B	Part C	Part D	Enthalpy change	Δ H° (a) at T_2 (J/mol)
25	298.15						
30	303.15	16.245	0.634372	0	-0.459151186	144.1524776	144.1524776
35	308.15	32.490	1.279293	0	-0.903402123	288.2687872	288.2687872
40	313.15	48.735	1.934765	0	-1.333466537	432.3548628	432.3548628
45	318.15	64.980	2.600786	0	-1.75001329	576.4162653	576.4162653
50	323.15	81.225	3.277358	0	-2.153669844	720.4582113	720.4582113
55	328.15	97.470	3.964479	0	-2.545025422	864.4855998	864.4855998
60	333.15	113.715	4.662151	0	-2.924633871	1008.503035	1008.503035
65	338.15	129.960	5.370372	0	-3.293016284	1152.51485	1152.51485
70	343.15	146.205	6.089144	0	-3.650663379	1296.525124	1296.525124
75	348.15	162.450	6.818465	0	-3.998037686	1440.537703	1440.537703
80	353.15	178.695	7.558337	0	-4.335575541	1584.556215	1584.556215
85	358.15	194.940	8.308758	0	-4.663688914	1728.584084	1728.584084
90	363.15	211.185	9.06973	0	-4.982767084	1872.624547	1872.624547
100	373.15	243.675	10.62332	0	-5.595270617	2160.755333	2160.755333
105	378.15	259.920	11.41594	0	-5.889374356	2304.851297	2304.851297
110	383.15	276.165	12.21912	0	-6.175802152	2448.971155	2448.971155
115	388.15	292.410	13.03284	0	-6.454850642	2593.117375	2593.117375
120	393.15	308.655	13.85711	0	-6.72680137	2737.292297	2737.292297
125	398.15	324.900	14.69193	0	-6.99192174	2881.498143	2881.498143
130	403.15	341.145	15.5373	0	-7.250465888	3025.737028	3025.737028
135	408.15	357.390	16.39322	0	-7.502675499	3170.01096	3170.01096
140	413.15	373.635	17.25969	0	-7.748780556	3314.321852	3314.321852
145	418.15	389.880	18.13672	0	-7.989000045	3458.671523	3458.671523
150	423.15	406.125	19.02429	0	-8.2235426	3603.061709	3603.061709
155	428.15	422.370	19.92241	0	-8.452607107	3747.494064	3747.494064
160	433.15	438.615	20.83108	0	-8.676383274	3891.970164	3891.970164
165	438.15	454.860	21.7503	0	-8.895052144	4036.491514	4036.491514
170	443.15	471.105	22.68007	0	-9.108786594	4181.059553	4181.059553
175	448.15	487.350	23.6204	0	-9.317751782	4325.675652	4325.675652
180	453.15	503.595	24.57127	0	-9.522105579	4470.341125	4470.341125
185	458.15	519.840	25.53269	0	-9.721998964	4615.057227	4615.057227
190	463.15	536.085	26.50466	0	-9.917576395	4759.825159	4759.825159
195	468.15	552.330	27.48718	0	-10.10897616	4904.646071	4904.646071
200	473.15	568.575	28.48025	0	-10.2963307	5049.521063	5049.521063
205	478.15	584.820	29.48387	0	-10.47976692	5194.451191	5194.451191
210	483.15	601.065	30.49805	0	-10.65940647	5339.437466	5339.437466
215	488.15	617.310	31.52277	0	-10.83536601	5484.480858	5484.480858
220	493.15	633.555	32.55804	0	-11.00775748	5629.582298	5629.582298
225	498.15	649.800	33.60386	0	-11.17668831	5774.742679	5774.742679
230	503.15	666.045	34.66023	0	-11.34226168	5919.962858	5919.962858

Enthalpy calculations						
Δ H°(b) at T_2 :	[Δ H° (at T_1=298.15°K)] + R*[A*(T_2-T_1) + B/2*(T_2^2-T_1^2)+C/3*(T_2^3-T_1^3) - D(1/T_2-1/T_1)]					
Gas Constant	R =		8.314	J/mol.K		
Operating Temperature	For Gas (b) or O_2 in gaseous form					
°C	°K	Part A	Part B	Part C	Part D	Enthalpy change	Δ H° (b) at T_2 (J/mol)
25	298.15						
30	303.15	18.195	0.760645	0	1.255750834	147.1569159	147.1569159
35	308.15	36.390	1.533939	0	2.470750383	294.7578102	294.7578102
40	313.15	54.585	2.319884	0	3.64695065	442.7864537	442.7864537
45	318.15	72.780	3.118478	0	4.786180925	591.2276379	591.2276379
50	323.15	90.975	3.929723	0	5.890157285	740.0670952	740.0670952
55	328.15	109.170	4.753617	0	6.960491214	889.2914278	889.2914278
60	333.15	127.365	5.590162	0	7.998697456	1038.888042	1038.888042
65	338.15	145.560	6.439356	0	9.006201161	1188.845089	1188.845089
70	343.15	163.755	7.301201	0	9.984344421	1339.151411	1339.151411
75	348.15	181.950	8.175695	0	10.93439223	1489.796491	1489.796491
80	353.15	200.145	9.06284	0	11.85753793	1640.770407	1640.770407
85	358.15	218.340	9.962634	0	12.75490823	1792.063792	1792.063792
90	363.15	236.535	10.87508	0	13.62756781	1943.667794	1943.667794
100	373.15	272.925	12.73792	0	15.30272807	2247.774615	2247.774615
105	378.15	291.120	13.68831	0	16.10708408	2400.262009	2400.262009
110	383.15	309.315	14.65136	0	16.89044685	2553.029113	2553.029113
115	388.15	327.510	15.62705	0	17.65362766	2706.069182	2706.069182
120	393.15	345.705	16.6154	0	18.39739652	2859.375814	2859.375814
125	398.15	363.900	17.61639	0	19.12248476	3012.942928	3012.942928
130	403.15	382.095	18.63003	0	19.82958743	3166.764747	3166.764747
135	408.15	400.290	19.65633	0	20.51936552	3320.835774	3320.835774
140	413.15	418.485	20.69527	0	21.19244803	3475.150781	3475.150781
145	418.15	436.680	21.74687	0	21.84943386	3629.704787	3629.704787
150	423.15	454.875	22.81111	0	22.49089362	3784.49305	3784.49305
155	428.15	473.070	23.88801	0	23.11737125	3939.511046	3939.511046
160	433.15	491.265	24.97755	0	23.72938558	4094.754461	4094.754461
165	438.15	509.460	26.07975	0	24.32743177	4250.219181	4250.219181
170	443.15	527.655	27.19459	0	24.91198261	4405.901272	4405.901272
175	448.15	545.850	28.32209	0	25.48348981	4561.79698	4561.79698
180	453.15	564.045	29.46223	0	26.04238514	4717.902716	4717.902716
185	458.15	582.240	30.61502	0	26.5890815	4874.215046	4874.215046
190	463.15	600.435	31.78047	0	27.123974	5030.730685	5030.730685
195	468.15	618.630	32.95856	0	27.64744082	5187.44649	5187.44649
200	473.15	636.825	34.14931	0	28.15984421	5344.359448	5344.359448
205	478.15	655.020	35.3527	0	28.66153122	5501.466674	5501.466674
210	483.15	673.215	36.56875	0	29.15283456	5658.765402	5658.765402
215	488.15	691.410	37.79744	0	29.6340733	5816.252979	5816.252979
220	493.15	709.605	39.03879	0	30.10555358	5973.92686	5973.92686
225	498.15	727.800	40.29278	0	30.56756923	6131.784602	6131.784602
230	503.15	745.995	41.55942	0	31.02040242	6289.82386	6289.82386

Enthalpy calculations

Δ H°(c) at T$_2$:	[Δ H° (at T$_1$=298.15°K)] + R*[A*(T$_2$-T$_1$) + B/2*(T$_2^2$-T$_1^2$)+C/3*(T$_2^3$-T$_1^3$) - D(1/T$_2$-1/T$_1$)]					
Gas Constant	R =		8.314	J/mol.K		
Operating Temperature	For Gas (c) or H$_2$O in gaseous form					
°C	°K	Part A	Part B	Part C	Part D	Enthalpy change	Δ H° (c) at T$_2$ (J/mol)
25	298.15						
30	303.15	17.350	2.179713	0	-0.669364982	167.9351302	-241650.0649
35	308.15	34.700	4.395675	0	-0.647642932	330.4259453	-241487.5741
40	313.15	52.050	6.647888	0	-0.626961375	493.2267936	-241324.7732
45	318.15	69.400	8.93635	0	-0.607254904	656.3371312	-241161.6629
50	323.15	86.750	11.26106	0	-0.58846317	819.7564564	-240998.2435
55	328.15	104.100	13.62203	0	-0.57053042	983.4843058	-240834.5157
60	333.15	121.450	16.01924	0	-0.553405089	1147.52025	-240670.4797
65	338.15	138.800	18.4527	0	-0.53703942	1311.863894	-240506.1361
70	343.15	156.150	20.92241	0	-0.521389138	1476.514867	-240341.4851
75	348.15	173.500	23.42838	0	-0.506413147	1641.472829	-240176.5272
80	353.15	190.850	25.97059	0	-0.492073259	1806.737462	-240011.2625
85	358.15	208.200	28.54905	0	-0.478333953	1972.30847	-239845.6915
90	363.15	225.550	31.16376	0	-0.465162152	2138.18558	-239679.8144
100	373.15	260.250	36.50194	0	-8.156960778	2535.01258	-239282.9874
105	378.15	277.600	39.2254	0	-0.428753644	2637.651033	-239180.349
110	383.15	294.950	41.98511	0	-0.417563414	2804.750148	-239013.2499
115	388.15	312.300	44.78108	0	-0.40680563	2972.15424	-238845.8458
120	393.15	329.650	47.61329	0	-0.396458291	3139.863127	-238678.1369
125	398.15	347.000	50.48175	0	-0.38650078	3307.876637	-238510.1234
130	403.15	364.350	53.38646	0	-0.376913758	3476.19461	-238341.8054
135	408.15	381.700	56.32743	0	-0.367679071	3644.816895	-238173.1831
140	413.15	399.050	59.30464	0	-0.358779662	3813.74335	-238004.2566
145	418.15	416.400	62.3181	0	-0.350199496	3982.973842	-237835.0262
150	423.15	433.750	65.36781	0	-0.341923483	4152.508245	-237665.4918
155	428.15	451.100	68.45378	0	-0.333937415	4322.346441	-237495.6536
160	433.15	468.450	71.57599	0	-0.326227905	4492.488319	-237325.5117
165	438.15	485.800	74.73445	0	-0.318782329	4662.933774	-237155.0662
170	443.15	503.150	77.92916	0	-0.311588776	4833.682706	-236984.3173
175	448.15	520.500	81.16013	0	-0.304635997	5004.735023	-236813.265
180	453.15	537.850	84.42734	0	-0.297913367	5176.090636	-236641.9094
185	458.15	555.200	87.7308	0	-0.291410838	5347.749461	-236470.2505
190	463.15	572.550	91.07051	0	-0.285118906	5519.71142	-236298.2886
195	468.15	589.900	94.44648	0	-0.279028573	5691.976437	-236126.0236
200	473.15	607.250	97.85869	0	-0.273131319	5864.544442	-235953.4556
205	478.15	624.600	101.3072	0	-0.267419068	6037.415367	-235780.5846
210	483.15	641.950	104.7919	0	-0.26188416	6210.58915	-235607.4109
215	488.15	659.300	108.3128	0	-0.25651933	6384.065729	-235433.9343
220	493.15	676.650	111.87	0	-0.251317681	6557.845047	-235260.155
225	498.15	694.000	115.4635	0	-0.246272661	6731.92705	-235086.073
230	503.15	711.350	119.0932	0	-0.241378044	6906.311686	-234911.6883

Enthalpy calculations

ΔH°(d) at T$_2$:	[ΔH° (at T$_1$=298.15°K)] + R*[A*(T$_2$-T$_1$) + B/2*(T$_2^2$-T$_1^2$)+C/3*(T$_2^3$-T$_1^3$) - D(1/T$_2$-1/T$_1$)]					
Gas Constant	R =		8.314	J/mol.K		
Operating Temperature	For Gas (d) or H$_2$O in liquid form					
°C	°K	Part A	Part B	Part C	Part D	Enthalpy change	ΔH° (d) at T$_2$ (J/mol)
25	298.15						
30	303.15	43.560	1.879063	-0.08135326	0	377.1039947	-285452.896
35	308.15	87.120	3.789375	-0.16543486	0	754.4451183	-285075.5549
40	313.15	130.680	5.730938	-0.25228982	0	1132.022997	-284697.977
45	318.15	174.240	7.70375	-0.34196312	0	1509.837256	-284320.1627
50	323.15	217.800	9.707813	-0.43449978	0	1887.887522	-283942.1125
55	328.15	261.360	11.74313	-0.52994478	0	2266.17342	-283563.8266
60	333.15	304.920	13.80969	-0.62834314	0	2644.694577	-283185.3054
65	338.15	348.480	15.9075	-0.72973984	0	3023.450618	-282806.5494
70	343.15	392.040	18.03656	-0.8341799	0	3402.441169	-282427.5588
75	348.15	435.600	20.19688	-0.9417083	0	3781.665856	-282048.3341
80	353.15	479.160	22.38844	-1.05237006	0	4161.124305	-281668.8757
85	358.15	522.720	24.61125	-1.16621016	0	4540.816141	-281289.1839
90	363.15	566.280	26.86531	-1.28327362	0	4920.740991	-280909.259
100	373.15	653.400	31.46719	-1.52725058	0	5681.288236	-280148.7118
105	378.15	696.960	33.815	-1.65425408	0	6061.909882	-279768.0901
110	383.15	740.520	36.19406	-1.78466094	0	6442.763045	-279387.237
115	388.15	784.080	38.60438	-1.91851614	0	6823.847351	-279006.1526
120	393.15	827.640	41.04594	-2.0558647	0	7205.162425	-278624.8376
125	398.15	871.200	43.51875	-2.19675161	0	7586.707895	-278243.2921
130	403.15	914.760	46.02281	-2.34122186	0	7968.483385	-277861.5166
135	408.15	958.320	48.55813	-2.48932047	0	8350.488521	-277479.5115
140	413.15	1001.880	51.12469	-2.64109242	0	8732.722929	-277097.2771
145	418.15	1045.440	53.7225	-2.79658273	0	9115.186236	-276714.8138
150	423.15	1089.000	56.35156	-2.95583638	0	9497.878067	-276332.1219
155	428.15	1132.560	59.01188	-3.11889839	0	9880.798048	-275949.202
160	433.15	1176.120	61.70344	-3.28581374	0	10263.9458	-275566.0542
165	438.15	1219.680	64.42625	-3.45662745	0	10647.32096	-275182.679
170	443.15	1263.240	67.18031	-3.6313845	0	11030.92315	-274799.0769
175	448.15	1306.800	69.96563	-3.81012991	0	11414.75199	-274415.248
180	453.15	1350.360	72.78219	-3.99290866	0	11798.8071	-274031.1929
185	458.15	1393.920	75.63	-4.17976577	0	12183.08813	-273646.9119
190	463.15	1437.480	78.50906	-4.37074622	0	12567.59468	-273262.4053
195	468.15	1481.040	81.41938	-4.56589503	0	12952.32639	-272877.6736
200	473.15	1524.600	84.36094	-4.76525718	0	13337.28289	-272492.7171
205	478.15	1568.160	87.33375	-4.96887769	0	13722.46379	-272107.5362
210	483.15	1611.720	90.33781	-5.17680154	0	14107.86873	-271722.1313
215	488.15	1655.280	93.37313	-5.38907375	0	14493.49732	-271336.5027
220	493.15	1698.840	96.43969	-5.6057393	0	14879.34921	-270950.6508
225	498.15	1742.400	99.5375	-5.82684321	0	15265.424	-270564.576
230	503.15	1785.960	102.6666	-6.05243047	0	15651.72133	-270178.2787

| Fuel cell reaction enthalpy | $\Delta H°_{rxn}$ at T_2 = $\Delta H°(H_2O)$ at T_2 - $\Delta H°(H_2)$ at T_2 - 0.5*$\Delta H°(O_2)$ at T_2 |

Fuel cell reaction enthalpy is derived from the calculated values shown in Table 9 to Table 12, using the relation shown above and is plotted below (Figures 3 and 4):

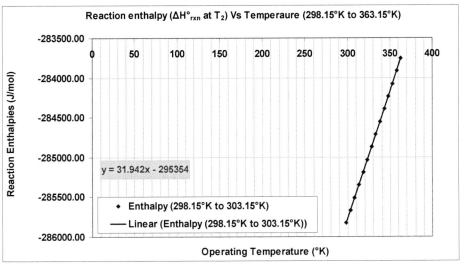

Appendix I – Figure 3

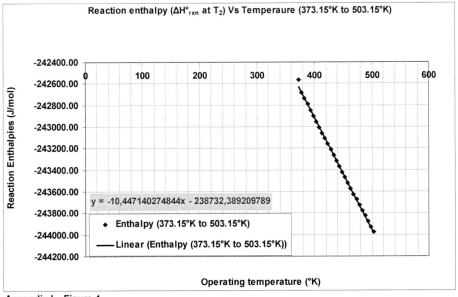

Appendix I – Figure 4

Conclusion: PEM fuel cell's reaction enthalpy ($\Delta H°_{rxn}$) at T_2 (J/mol)

At (25°C to 90°C or 298.15°K to 363.15°K)   = 31.942*T - 295354 (Temp in Kelvin)

At (100°C to 230°C or 373.15°K to 503.15°K) = -10.447140274844*T – 238732.3892097890

From the derived values of reaction entropy and enthalpy, Gibbs free energy of fuel cell reaction at various operating temperatures were derived and are shown in Figures 5 & 6:

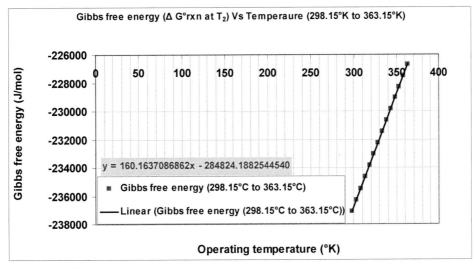

Appendix I – Figure 5

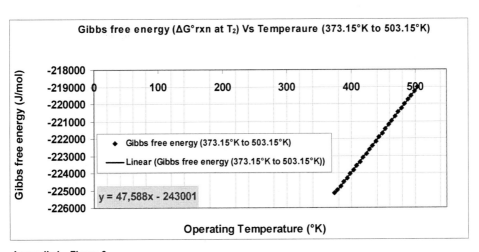

Appendix I – Figure 6

Conclusion: Gibbs free energy of fuel cell reaction ($\Delta G°_{rxn}$ at T_2) (J/mol)

At (25°C to 90°C or 298.15°K to 363.15°K)   160.1637086862*T − 284824.1882544540

At (100°C to 230°C or 373.15°K to 503.15°K) = 47.588*T − 243000.5744 (Temp in Kelvin)

Temp	Temp	Δ S°rxn at T2	Δ S°rxn/nF at T2	Δ H°rxn at T2	Δ G°rxn at T2	Maximum Therm. Efficiency at T_2	Max. Cell EMF at T_2 (Δ G°rxn at T2/ nF)
°C	°K	J/mol.K		J/mol	J/mol	Number	(V)
25	298.15	-163.430	-8.469E-04	-285830.00	-237103.35	0.829526	1.228706
30	303.15	-162.900	-8.442E-04	-285670.63	-236287.52	0.826672	1.224478
35	308.15	-162.378	-8.415E-04	-285511.20	-235474.33	0.823827	1.220264
40	313.15	-161.865	-8.388E-04	-285351.73	-234663.73	0.820991	1.216063
45	318.15	-161.359	-8.362E-04	-285192.19	-233855.67	0.818163	1.211876
50	323.15	-160.862	-8.336E-04	-285032.60	-233050.12	0.815345	1.207701
55	328.15	-160.372	-8.311E-04	-284872.96	-232247.04	0.812536	1.203540
60	333.15	-159.889	-8.286E-04	-284713.25	-231446.39	0.809734	1.199391
65	338.15	-159.413	-8.261E-04	-284553.49	-230648.14	0.806942	1.195254
70	343.15	-158.943	-8.237E-04	-284393.66	-229852.26	0.804157	1.191129
75	348.15	-158.481	-8.213E-04	-284233.77	-229058.70	0.801381	1.187017
80	353.15	-158.025	-8.189E-04	-284073.82	-228267.44	0.798613	1.182917
85	358.15	-157.575	-8.166E-04	-283913.80	-227478.44	0.795852	1.178828
90	363.15	-157.131	-8.143E-04	-283753.72	-226691.68	0.793100	1.174751
100	373.15	-46.670	-2.418E-04	-242567.63	-225152.86	0.787716	1.166777
105	378.15	-46.802	-2.425E-04	-242685.33	-224987.00	0.787136	1.165917
110	383.15	-46.933	-2.432E-04	-242738.74	-224756.22	0.786328	1.164721
115	388.15	-47.062	-2.439E-04	-242792.00	-224524.70	0.785518	1.163521
120	393.15	-47.190	-2.445E-04	-242845.12	-224292.45	0.784706	1.162318
125	398.15	-47.315	-2.452E-04	-242898.09	-224059.49	0.783891	1.161110
130	403.15	-47.439	-2.458E-04	-242950.92	-223825.81	0.783073	1.159900
135	408.15	-47.561	-2.465E-04	-243003.61	-223591.44	0.782253	1.158685
140	413.15	-47.682	-2.471E-04	-243056.15	-223356.39	0.781431	1.157467
145	418.15	-47.801	-2.477E-04	-243108.55	-223120.67	0.780606	1.156245
150	423.15	-47.918	-2.483E-04	-243160.80	-222884.28	0.779779	1.155020
155	428.15	-48.034	-2.489E-04	-243212.90	-222647.25	0.778950	1.153792
160	433.15	-48.148	-2.495E-04	-243264.86	-222409.57	0.778118	1.152560
165	438.15	-48.261	-2.501E-04	-243316.67	-222171.26	0.777285	1.151325
170	443.15	-48.372	-2.507E-04	-243368.33	-221932.32	0.776449	1.150087
175	448.15	-48.482	-2.512E-04	-243419.84	-221692.78	0.775611	1.148846
180	453.15	-48.590	-2.518E-04	-243471.20	-221452.63	0.774770	1.147601
185	458.15	-48.697	-2.524E-04	-243522.42	-221211.89	0.773928	1.146354
190	463.15	-48.803	-2.529E-04	-243573.48	-220970.57	0.773084	1.145103
195	468.15	-48.907	-2.534E-04	-243624.39	-220728.66	0.772238	1.143850
200	473.15	-49.010	-2.540E-04	-243675.16	-220486.19	0.771389	1.142593
205	478.15	-49.111	-2.545E-04	-243725.77	-220243.16	0.770539	1.141334
210	483.15	-49.212	-2.550E-04	-243776.23	-219999.57	0.769687	1.140071
215	488.15	-49.311	-2.555E-04	-243826.54	-219755.44	0.768833	1.138806
220	493.15	-49.409	-2.560E-04	-243876.70	-219510.78	0.767977	1.137538
225	498.15	-49.505	-2.565E-04	-243926.71	-219265.58	0.767119	1.136268
230	503.15	-49.601	-2.570E-04	-243976.56	-219019.86	0.766259	1.134994

Appendix II

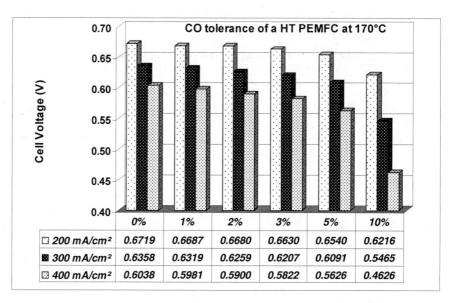

	0%	1%	2%	3%	5%	10%
☐ 200 mA/cm²	0.6719	0.6687	0.6680	0.6630	0.6540	0.6216
▦ 300 mA/cm²	0.6358	0.6319	0.6259	0.6207	0.6091	0.5465
▨ 400 mA/cm²	0.6038	0.5981	0.5900	0.5822	0.5626	0.4626

Appendix II – Figure 1: CO tolerance of a HT PEMFC at 170°C (H_2+CO/Air: 1.35/2.5)

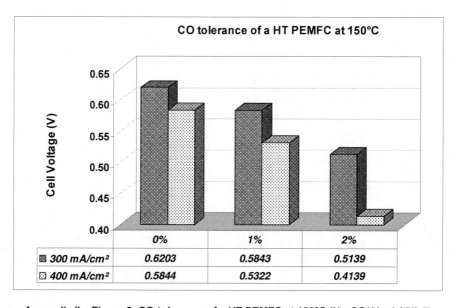

	0%	1%	2%
▦ 300 mA/cm²	0.6203	0.5843	0.5139
▨ 400 mA/cm²	0.5844	0.5322	0.4139

Appendix II – Figure 2: CO tolerance of a HT PEMFC at 150°C (H_2+CO/Air: 1.35/2.5)

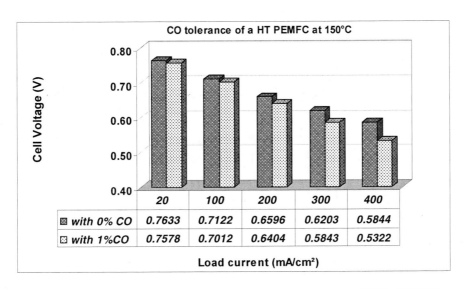

	20	100	200	300	400
with 0% CO	0.7633	0.7122	0.6596	0.6203	0.5844
with 1%CO	0.7578	0.7012	0.6404	0.5843	0.5322

Load current (mA/cm²)

Appendix II – Figure 3: CO tolerance of a HT PEMFC at 150°C (H₂+CO/Air: 1.35/2.5)

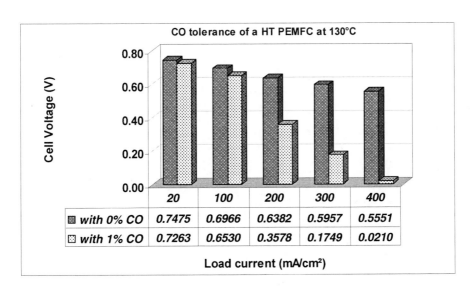

	20	100	200	300	400
with 0% CO	0.7475	0.6966	0.6382	0.5957	0.5551
with 1% CO	0.7263	0.6530	0.3578	0.1749	0.0210

Load current (mA/cm²)

Appendix II – Figure 4: CO tolerance of a HT PEMFC at 130°C (H₂+CO/Air: 1.35/2.5)

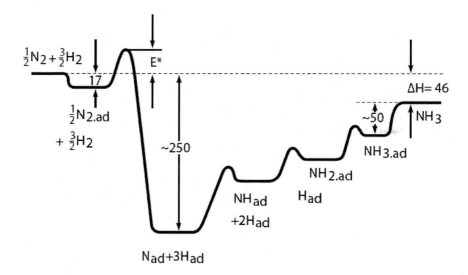

An energy diagram showing the progression of the reaction from the reactants N_2 and H_2 to the product NH_3. Energies are given in units of kJ/mol. (Adapted from Ertl 1983)

$$H_2 \rightleftharpoons 2\ H_{ad}$$

$$N_2 \rightleftharpoons N_{2,ad} \rightleftharpoons 2\ N_s$$

$$N_s + H_{ad} \rightleftharpoons NH_{ad}$$

$$NH_{ad} + H_{ad} \rightleftharpoons NH_{2,ad}$$

$$NH_{2,ad} + H_{ad} \rightleftharpoons NH_{3,ad}$$

$$NH_{3,ad} \rightleftharpoons NH_3$$

Appendix II – Figure 5: Formation of Ammonia from H_2 and N_2 (at temperatures > 400°C)

Appendix II – Figure 6:   Fuel cell stack (HT PEMFC) equipped Antares (at Hamburg airport, Germany), 2009. (Source: DLR, Germany).

Appendix II – Figure 7: Antares test flight (with HT PEMFC stack containing BASF's Celtec ® P based MEAs and a stack built by SerEnergy, Denmark), from Hamburg Airport, Germany. (Source: DLR, Germany)

Antares DLR-H2 specifications	Technical data
Span of the Flight	20 m / 65,6 ft
Flight's body area	12.6 m² / 13 ft²
Longer main body length:	7.4 m / 24.3 ft
Long POD:	2.87 m / 9.43 ft
Diameter POD:	0.6 m / 1.97 ft
Weight without any FC system:	Approx. 460 kg / 1014 lb.
Weight of the Fuel cell stack:	Approx. 60 kg
Fuel cell stack's body temp – in flight:	<-45° C to 40° C
Maximum allowed load of DLRH2:	750 kg (>900 kg in 4 POD-Version)
Range:	>750 km (>2.000 km in 4 POD-Version)
Max. Fuel cell stack power:	~ 25 kW (up to 45 kW in 4 POD-Version)
Fuel cell stack continuous power:	> 20 kW
Needed power for take off:	Approx. 10 kW
Max. speed:	Approx. 170 km/h
Speed to be buoyant:	Up to 300 km/h
Max. reachable height:	>> 4000 m / >> 12.000 ft
Max. lowering abilitiy (560kg):	>2.5 m/s (at 25 kW)
Total drive train efficieny:	Approx. 44 Percent

References

[1]   Lecture notes of Prof.Jeff Tester, MIT, USA, as documented by Yang Shao-Horn, Electrochemical Energy Laboratory, MIT, USA, 2004.

[2]   Hrad Paul.M, Thesis, Air Force Institute of Technology, AFIT/GAE/ENY/10-M12, March 2010.

[3]   Gervasio, D; Kinder J, Hoskins N, High temperature polymer electrolyte membrane fuel cells for portable power, Journal of the Society of Automotive Engineers (SAE): 2006-01-3096, 2006. USA.

[4]   James D.B, Kalinoski J.A, Mass production cost estimation for direct H_2 PEM fuel cell systems for automotive applications – 2008 update, March 2009, v.30.2021.052209. Report of Directed technologies, Virginia, USA.

[5]   Costamagna P, Srinivasan S, J. Power Sources 102 (2001), p. 242.

[6]   Vogel J, Development of Polybenzimidazole based high temperature membrane and electrode assemblies for stationary and automotive applications, Final report: DE-FC36-03G013101, US DoE, 2008.

[7]   Cost analysis of fuel cell systems for transportation: baseline system cost estimate. SFAA DE-SCO2–98EE50526, Final report, US DOE, Washington, DC. USA.

[8]   Brandon N.P, Skinner S, Steele B.C.H, Recent advances in materials for fuel cells, Annu. Rev. Mater. Res. 2003. 33:183–213, doi: 10.1146/annurev.matsci.33.022802.094122.

[9]   Carlson E.J, Yang Y, Fulton C, SOFC manufacturing cost model: Simulating relationships between performance, manufacturing and cost of production, DoE report, 2004.

[10] Aravind PV, Carlson E.J, Yang Y, Fulton C, SOFC manufacturing cost model: Simulating relationships between performance, manufacturing and cost of production, DoE report, 2004.

[11] Kerres J, State of art of membrane development, J.Membr.Sci 185 (2001) 1.

[12] Meier-Haack J, Taeger A, Vogel C, Schlenstedt K, Lenk W, Lehmann D, Membranes from sulfonated block copolymers for use in fuel cells, Sep.Purification Technol. 41(3) (2005) 207 – 220.

[13] Smitha B, Sridhar S, Khan A.A, Solid polymer electrolyte membranes for fuel cell applications - a review, J.Membr.Sci 259 (2005) 10-26.

[14] Mauritz K.A, Warren R.M, 1989. Macromolecules 22; 1730 - 34

[15] Alberti G, Casciola M, Composite membranes for Medium temperature PEM fuel cells, Annual review of materials research, 2003. 33: 129-54.

[16] Roziere J, Jones D.J, Non-fluorinated polymer materials for proton exchange membrane fuel cells, Annu. Rev. Mater. Res. (2003) 33, 503 – 555.

[17] Jones D.J, Roziere J, Recent advances in the functionalisation of polybenzimidazole and polyetherketone for fuel cell applications, J. Membr. Sci. 185 (2001) 41–58.

[18] Kreuer K.D, On the development of proton conducting polymer membranes for hydrogen and methanol fuel cells. J.Membr.Sci. 185 (2001) 29-39.

[19] Singleton R.W, Noether H.D, Tracy J.F., 1967. J.Polym.Sci.Polym.Symp.19:65-76.

[20] Buckley A, Stuetz D.E, Serad G.A, Encyclopedia of Polymer Science and Engineering, ed. HF Mark, New York; Wiley & Sons, 2nd ed, 11: 572-601, 1982.

[21] Gillham J.K., Crit. Rev. 1972, Macromol. Sci. 1:83.

[22] Donoso P, Gorecki W, Berthier C, Defendini F, Poinsignon C, Armand M.B, 1988. Solid State Ionics 28:969-74.

[23] Daniel M.F, Desbat B, Cruege F, Trinquet O, Lassegues J.C, 1988. Solid State Ionics 28:637-41.

[24] Schuster, Martin F.H, Meyer, Wolfgang H, Anhydrous proton conducting polymers, Annual review of materials research, 2003. 33: 233-61.

[25] Wainright J.S, Wang J.T, Weng D, Savinell R.F, Litt M, 1995, J. Electrochem.Soc.142(7):121-23.

[26] Herz H.G, Kreuer K.D, Maier J, Scharfenberger G, Schuster M.F.H, Meyer W.H, Electrochem.Acta. 2003, 48, 2165.

[27] Schechter A, Savinell RF. 2002. Solid State Ionics 147(1-2):181-87.

[28] Dalmia A, Liu C.C, Savinell, R.F, J.Electroanal. Chem. 1997, 430, 205 – 214.

[29] Wainright J.S, Wang J.T, Savinell R.F, Litt M, Moaddel H, Rogers C, J.Electrochem Soc. 1994, 23 (255-264).

[30] Samms S.R, Savinell R.F, J.Electrochem.Soc. 1996, 143, 1225.

[31] Weng D, Landau U, Savinell R.F, J.Electrochem. Soc. 1996, 143, 1260.

[32] Savinell R.F, Litt M.H, 1996, WO Patent 9613872.

[33] Li Q, Hjuler H.A, Bjerrum N.J, Phosphoric acid doped Polybenzimidazole membranes: physiochemical characterization and fuel cell applications, J. Appl.Electrochem 31(7):773-779, 2001.

[34] U.S.Patent 6946211B1, Polymer electrolyte membrane fuel cells, Sep.20, 2005.

[35] Xiao L, PhD dissertation, Rensselaer Polytechnic Institute, Troy, New York, 2003.

[36] Smith, Van Ness and Abbott, "Chemical Engineering Thermodynamics, 2000"

[37] Chase M.W, "JANAF Thermochemical Tables", 3rd Edition, American Chemical Society and the American Institute of Physics for the national Bureau of Standards (or National Institute of Standards and Technology), 1985.

[38] Nernst W, Ueber die Berechnung chemischer Gleichgewichte aus thermischen Messungen. Nachr. Kgl. Ges. Wiss. Gött., 1906, No. 1, pp. 1 – 40.

[39] Appleby J, and Foulkes F, "Fuel Cell Handbook", Texas A&M University, Van Nostrand Reinhold, New York, republished by Krieger Publishing Co., Melbourne, FL, USA,1989 and "Fuel Cell Handbook", EG&G services, US DoE, 2000.

[40] Bard A. J, Faulkner L. R, Electrochemical Methods. Fundamentals and Applications, 2nd Ed. Wiley, New York. 2001. ISBN 0-471-04372-9.

[41] Fick A, Phil. Mag. (1855), 10, 30.

[42] Springer T.E, Wilson M.S, Gottesfeld S, Modeling and Experimental Diagnostics in Polymer Electrolyte Fuel Cells", J.Electrochem. Soc. 140, 3513 (1993).

[43] Perry M.L, Newman J, Cairns E.J, Mass transport in Gas-Diffusion Electrodes: A Diagnostic Tool for Fuel-Cells Cathodes, J. Electrochem. Soc. 145, 5 (1998).

[44] Gasteiger H.A, Mathias M.F, Research Paper: Fundamental research and development challenges in polymer electrolyte fuel cell technology, General Motors corporation, Honeoye Falls, NY, USA, 2004.

[45] Vielstich W, Gasteiger H.A, Yokokawa, Handbook of fuel cells, Volume 5, Advances in electrocatalysis, materials, diagnostics and durability, Part I, ISBN: 978-0-470-72311-1.

[46] Cheddie D.F, Computational modelling of intermediate temperature proton exchange membrane fuel cells, PhD Thesis, 2006.

[47] Savadogo O, Xing B, Hydrogen-oxygen polymer electrolyte membrane fuel cell (PEMFC) based on acid-doped polybenzimidazole (PBI), J. New Materials for Electrochemical Systems 3 (2000) 345.

[48] Shamardina O, Chertovich A, Kulikovsky A.A, Khokhlov A.R, A simple model of a high

temperature PEM fuel cell, International Journal of Hydrogen Energy (2009), doi:10.1016/j.ijhydene.2009.11.012.

[49]   Wang J.T, Savinell R.F, Wainright J, Litt M, Yu H, A H_2/O_2 fuel cell using acid doped Polybenzimidazole as polymer electrolyte, Electrochemica Acta 41 (1996) 193.

[50]   Broka K, Techn.Lic, Thesis, Royal Institute of Technology, Stockholm, 1995.

[51]   Scott K, Pilditch S, Mamlouk M, Modelling and experimental validation of a high temperature polymer electrolyte fuel cell, J.Appl.Electrochem (2007) 37:1245-1259. DOI: 10.1007/s10800-007-9414-1.

[52]   Song Chaojie, Hui Rob and Zhang Jiujun, PEM fuel cell electrocatalysts and catalyst layers: fundamentals and applications, ISBN: 978-1-84800-935-6; DOI: 10.1007/978-1-84800-936-3. pp 871 – 884.

[53]   Li Q, Jensen J.O, Noye P.P, Pan C and Bjerrum N.J, Proton conductivity and operational features of PBI-based membranes, Proceedings of the 26th Riso International Symposium on Materials Science: Solid state electrochemistry, Riso National laboratory, Roskilde, Denmark, 2005.

[54]   Broka K, Ekdunge P, Oxygen and hydrogen permeation properties and water uptake of Nafion 117 membrane and recast film for PEM fuel cell. J Appl.Electrochem. 1997 (27): 117-123.

[55]   Mecerreyes D, Grande H, Miguel O, Ochoteco E, Marcilla R, Cantero I, "Porous polybenzimidazole membranes doped with phosphoric acid: Highly proton conducting solid electrolytes". Chem Mater 2004 (16): 604-607.

[56]   Kumbharkar S.C, Karadkar P.B, Kharul U.K, Enhancement of gas permeation properties of polybenzimidazoles by systematic structure architecture, J. Membr. Sci. 286 (2006), pp.161.

[57]   He R.H, Li Q.F, Bach A, Jensen J.O, Bjerrum N.J, Physicochemical properties of phosphoric acid doped polybenzimidazole membranes for fuel cells. J Membr. Sci. 277 (2006) :38-45.

[58]   Li Qingfeng, Jensen J.O, Savinell R, Bjerrum N.J, High temperature proton exchange membranes based on polybenzimidazoles for fuel cells, Progress in polymer science (2008), doi:10.1016/j.progpolymsci.2008.12.003.

[59]   Neyerlin K.C, Singh A, Chu D, Kinetic characterization of a Pt–Ni/C catalyst with a phosphoric acid doped PBI membrane in a proton exchange membrane fuel cell, Journal of Power Sources 2008 (176):112–117.

[60]   Liu Z, Wainright J.S, Litt M.H, Savinell R.F, Study of the oxygen reduction reaction (ORR) at Pt interfaced with phosphoric acid doped polybenzimidazole at elevated temperature and low relative humidity, Electrochem Acta 51 (2006) 3914.

[61] Klinedinst K, Bett J.A.S, Macdonald J, Stonehart P, Oxygen solubility and diffusivity in hot concentrated H_3PO_4, Electroanalytical Chemistry and Interfacial Electrochemistry, 57 (1974) 281 – 281.

[62] Bindra P, Clouser S.J, Yeager E, J.Electrochem.Soc. 1979 (126), 1631.

[63] Aragane J, Murahashi T, Odaka T, J. Electrochem. Soc. 1988 (135), 844.

[64] Pourbaix M, Atlas of Electrochemical equilibria in aqueous solutions, National association of corrosion engineers, New York, 1974, pp 379.

[65] Liu G, Zhang H, Zhai Y, Zhang Y, Xu D, Shao Z, Pt_4ZrO_4/C cathode catalyst for improved durability in high temperature PEMFC based on H_3PO_4 doped PBI, Electrochem Commun (9): 135-41, (2007).

[66] Wang X, Waje M, Yan Y, CNT based electrodes with high efficiency for PEMFCs, Electrochem Solid state letters, 2005 (8): A42-4.

[67] Shim J, Lee C.R, Lee H.K, Lee J.S, Cairns E.J, Electrochemical characteristics of Pt-WO_3/C and Pt-TiO_2/C electrocatalysts in a polymer electrolyte fuel cell, J Power Sources, 2001; (102): 172-7.

[68] He Qinggang, Yang X, Chen W, Mukerjee S, Koel B, Chen S, Influence of phosphate anion adsorption on kinetics of oxygen electro reduction on low index Pt (hkl) single crystals, Phys. Chem.Chem.Phys., 2010, 12, 12544-12555.

[69] Hoel D, Grunwald E, J.Phys.Chem 81, 2135 (1977).

[70] Li Q, He R, Jensen J. and Bjerrum N (2004), PBI-Based Polymer Membranes for High Temperature Fuel Cells – Preparation, Characterization and Fuel Cell Demonstration. Fuel Cells, 4: 147–159. doi: 10.1002/fuce.200400020.

[71] Bouchet R, Siebert E, Proton conduction in acid doped Polybenzimidazole, Solid state Ionics, 1999, 118: 287-299.

[72] Evertz J, Günthart, report from Tribecraft AG, Switzerland.

[73] Peng J, Shin J.Y, Song T.W, Transient response of a high temperature PEM fuel cell, Journal of power sources, 179 (2008): 220 -231.

[74] Barsoukov E, Macdonald J.R, Impedance spectroscopy: Theory, Experiment and Applications, ISBN: 0-471-64749-7, John Wiley & Sons Inc, 2005.

[75] Orazem M.E, Tribollet B, Electrochemical Impedance Spectroscopy, ISBN: 978-0-470-04140-6, John Wiley & Sons Inc, 2008.

[76] Lasia, Andrzej, Electrochemical Impedance Spectroscopy and its Application: Modern

Aspects of Electrochemistry, 32, Conway B.E, Bockris J.O'M, Whitre R.E, et al., Kluwer Academic/Plenum Publishers, New York, (1997). ISBN: 0-306-45964-7.

[77] Sluyters-Rehbach M, Impedances of Electrochemical Systems : Terminology , Nomenclature and Representation Part I: Cells with Metal Electrodes and Liquid Solution, Pure & Applied Chemistry, 66 (9), (1994), 1831-1891.

[78] Sadkowski A, Time Domain Responses of Constant Phase Electrodes, Electrochimica Acta, 38 (14), (1993), 2051-2054.

[79] Wang J.C, Impedance of a Fractal Electrolyte-Electrode Interface, Electrochimica Acta, 33 (5), (1988), 707-711.

[80] Brug G.L, Van Den Eeden A.L.G, Sluyters-Rehbach M, and Sluyters J.H, The Analysis of Electrode Impedances Complicated by the Presence of a Constant Phase Element, Journal of Electroanaytical.Chemistry., 176, (1984), 275-295.

[81] Van Heuveln, Fred H., Analysis of Nonexponential Transient Response Due to a Constant Phase Element, Journal of the Electrochemical Society, 141(12), (1994), 3423-3428.

[82] Zoltowski, Piotr, On the electrical capacitance of interfaces exhibiting constant phase element behaviour, Journal of Electroanalytical Chemistry, 443, (1998), 149-154.

[83] Wainright JS, Wang J.T, Weng D, Savinell R.F, Litt M, 1995. J. Electrochem. Soc.142:L121–23.

[84] Marianowski L.G, 160°C Proton Exchange Membrane Fuel Cell System Development, Prepared by Gas Technology Institute under DoE cooperative agreement No: DE-FC26-99FT40656, Dec, 2001.

[85] Xiao Lixiang, Zhang Haifend, Scanlon Eugene, Ramanathan L.S, Choe Eui-Won, Rogers D, Apple T and Benicewicz B, High temperature polybenzimidazole fuel cell membranes via Sol-Gel process, Chem. Mater. 2005, 17, 5328-5333.

[86] Li Q, He R, Jensen J. and Bjerrum N (2004), PBI-Based Polymer Membranes for High Temperature Fuel Cells – Preparation, Characterization and Fuel Cell Demonstration. Fuel Cells, 4: 147–159. doi: 10.1002/fuce.200400020.

[87] Li Q, Pan C, Jensen J.O, Noyé P, Bjerrum N.J, Cross-linked polybenzimidazole membranes for fuel cells. Mater. Chem. 2007, 19:350-352.

[88] Webpage: http://www.adventech.gr/products.html (visited in 2009).

[89] Bandlamudi George, Burfeind J, Siegel C, Heinzel A, Changes in membrane conductivity introduced by hydration, dehydration cycles of phosphoric acid, Proceedings of the Fuel cells science and technology 2008 conference, Copenhagen, Denmark.

[90] Behret H, Binder H, Sandstede G, Scherer G. G, J. Electroanal.Chem. 1981, (117) 29.

[91] Kamat Ashish, In-Situ characterization of the membrane electrode assembly of a HT-PEM Fuel Cell, Internal presentation at Volkswagen AG, Germany, September 2008.

[92] US Department of Energy Report, prepared by EG&G Services, West Virginia, October 2000.

[93] Weber A.Z, J. Electrochem. Soc., 155 (6) , B521-531 (2008).

[94] Wang J.T, et al., Electrochimica Acta, Vol. 41(2), 193-197 (1996).

[95] G. Liu, H. Zhang, J. Hu, Y. Zhai, D. Xu, Z. Shao, Studies of performance degradation of high temperature PEMFC based on H_3PO_4 doped PBI, J. Power Sources 162 (2006), 547-552.

[96] R.Staudt, DoE report on PBI based high temperature MEAs, May, 2005.

[97] Shamardina O, et al., A simple model of a high temperature PEM fuel cell, International Journal of Hydrogen Energy (2009), doi:10.1016/j.ijhydene.2009.11.

[98] Wainright J.S, M. H. Litt, R. F. Savinell, "High-temperature membranes in Handbook of Fuel Cells – Fundamentals, Technology and Applications", Vol. 3: "Fuel Cell Technology and Applications Part 1", edited by W. Vielstich, H.A. Gasteiger, A. Lamm, John Wiley & Sons (2003), 436-446.

[99] Gottesfeld Shimshon, The Polymer Electrolyte Fuel Cell: Materials Issues in a Hydrogen Fueled Power Source, Report, Los Alamos National Laboratory, 1990, USA.

[100] Tang Y, Zhang J, Song C, Zhang J, Single PEMFC Design and validation for high temperature MEA testing and diagnosis upto 300°C, Electrochemical and solid state letters, 10 (9) B142-B146 (2007).

[101] Bandlamudi George, Renewables and Fuel Cells for Air Pollution mitigation in India, Department of Energy and Semiconductor research, University of Oldenburg, Germany, September 2004.

[102] Koshimizu T, Lto K, Satomi T, Development of 5000 kW and 1000 kW PAFC plants, presented at the JASME-ASME joint Conference (ICOPE-93), Tokyo, Japan, 1993.

[103] James M.Douglas, conceptual Design of Chemical Processes, McGraw-Hill Inc., New York, USA, 1988. ISBN: 978-0070177628.

[104] Schmidt T.J, Baurmeister J, ECS Transactions, 3(1) 861-869 (2006).

[105] Lundberg W.L, Solid Oxide Fuel Cell Cogeneration System Conceptual Design, prepared by Westinghouse for Gas Research Institute, Report: RI-89-0162, July, 1989.

[106] Minh N.Q, High Temperature Fuel Cells, Part 2: The Solid Oxide Cell, Chemtech, Vol.21, February 1991.

[107] Ertl Gerhard, Chemical processes on solid surfaces – Scientific background on Nobel Prize in Chemistry, The Royal Swedish Academy of sciences, Sweden, 2007.

[108] Schmidt T.J, BASF, Germany, Durability and degradation in high temperature polymer electrolyte fuel cells, in our internal communications report, Germany, 2005.

[109] Personal communication with Dr.Isabel Kundler, BASF Fuel Cells, Germany, 2006.

[110] Lorenz Gubler, Kramer Denis, Belack Jörg, Ünsal Ömer, Schmidt Thomas.J, Scherer Günther.G, A Polybenzimidazole-based membrane for the direct methanol fuel cell, Journal of Electrochemical society, 154(9) B981-B987, 2007.

[111] G.Calundann, Carsten Henschel, MEAs for reformed hydrogen fuel cells, PEMEAS Fuel Cell Technologies (Currently BASF), Fuel Cell Seminar 2004, USA.

[112] Los Angeles Auto Show, 2007, http://www.autoindustry.co.uk/news.

[113] http://www.ultracellpower.com/sp.php?xx25.

[114] Kolb G, Schelhaas, Wichert M, Burfeind J, Heßke C, Bandlamudi G, Entwicklung eines mikrostrukturierten Methanolreformers gekoppelt mit einer Hochtemperatur-PEM Brennstoffzelle, 81, No.5, Chemie Ingenieur Technik, 2009, Germany; and http://www.basf-fuelcell.com/en/innovation-scouting/research-projects/micropower.html

[115] http://www.nextgencell.eu/NextGenCell%20Website.htm.

[116] Aicher Thomas, Post-graduate lecture course, Bio-energy – theory and applications, Helsinki University of Technology, November, 2005.

[117] Damle S. Ashok, Hydrogen production by reforming of liquid hydrocarbons in a membrane reactor for portable power generation, J.Power Sources, 186 (2009): 167 – 177.

[118] Agrell J, Boutonnet M, Fierro J.L.G, Production of hydrogen from methanol over binary Cu\ZnO catalysts; Part II. Catalytic activity and reaction pathways, Applied Catalysis; A: General 253 (2003): 213 -223; DOI: 10.1016/S0926-860X(03)00521-0.

[119] Kirillov V.A, Meshcheryakov V.D, Sobyanin V.A, Belyaev V.D, Amosov Yu.I, Kuzin N.A, Bobrin A.S, Bioethanol as a promising fuel for fuel cell power plants, Theoretical foundations of chemical engineering, 2008, Vol.42 (1), pp 1- 11, DOI: 10.1134/S0040579508010016.

[120] Spitta Christian, PhD Dissertation, University of Duisburg-Essen, Duisburg, 2008.

[121] Cairns, E.J, J.Electrochem.Soc, 113: 1200-1204, 1966.

[122] Savadogo O, Varela F.J.R, J. New Mater. Electrochemical systems, 2001 (4) : 93-97.

[123] Psofogiannakis G, Bourgault Y, Conway B.E and Ternan M, Mathematical model for a direct propane phosphoric acid fuel cell, J. Applied electrochemistry (2006) 36(1): 115 – 130. DOI: 10.1007/s10800-005-9044-4.

[124] Bandlamudi G.C, Saborni M, Beckhaus P, Mahlendorf F and Heinzel Angelika, PBI/H$_3$PO$_4$ gel based polymer electrolyte membrane fuel cells under the influence of reformates, J. Fuel Cell Sci Tech 7 (2010), pp.014501–014502.

[125] Smolinka, Wittstadt, Gruenerbel, Lehnert, Performance and endurance of PEFC single cells and stacks fed with hydrogen and reformate, Proceedings of the 3rd European PEFC forum, Lucerne, Switzerland, 2005.

[126] Hinds G, Performance and durability of PEM fuel cells: a review, ISSN-1744-0262, 2004.

[127] Yamazaki O, Oomori Y, Shintaku H, Tabata T, Proceedings of the Fuel Cell Seminar, Palm Springs, USA, 2005.

[128] Cleghorn S.J.C, Mayfield D.K, Moore D.A, Moore J.C, Rusch G, Sherman T.W, Sisofo N.T, Beuscher U, J.Power Sources 2006, 158, 446.

[129] St.Pierre J, Wilkinson D.P, Knights S.D, Bos M, J. New Mater. Electrochem. Sys 2000, 3, 99.

[130] Shi Y, Horky A, Polevaya O, Cross J, Abstracts of the Fuel Cell Seminar, 2005, Courtesy Associates, Palm Springs. USA, 2005.

[131] Endoh E, Proceedings of the Fuel Cell Seminar, Palm Springs, USA, 2005.

[132] Hou J, Yu H, Yi B, Xiao Y, Wang H, Yi B, Ming P, Electrochem.Solid-state Lett. 10, B11, 2007.

[133] Garche J, Alterungserscheinung an der MEA in PEMFC´s, WBZU, January 2007.

[134] Kundler I, Henschel C, BASF fuel cells (formerly PEMEAS), Communications in line with long term tests performed at ZBT Duisburg, 2005.

[135] Kaufman Lars, RWE energy AG, Dortmund, Communications in line with long term tests performed at ZBT Duisburg, 2005.

[136] Bandlamudi G, Heinzel A, Derieth T, Burfeind J, Buder I, Beckhaus P, Addressing component challenges when operating HT PEMFCs containing H$_3$PO$_4$ based electrolytes, Progress MEA (CARISMA), La Grande Motte, September, 2008, France.

[137] Calundann G, Henschel C, MEAs for reformed hydrogen fuel cells, PEMEAS Fuel Cell

Technologies (Currently BASF), Fuel Cell Seminar 2004, USA.

[138] Staudt, R, Development of Polybenzimidazole-based High Temperature Membrane and Electrode Assemblies for Stationary Applications, DOE Hydrogen Program 2006 Merit Review, Arlington, Virginia, USA, May 16-19, 2006.

[139] Schmidt Thomas, BASF fuel cells (formerly PEMEAS), communications in line with initial projects with Celtec P ® MEAs with ZBT Duisburg, Germany, 2005.

[140] Jansen, High temperature PEM Fuel cells - from membranes to stacks, Presentation at HySA Systems Mini-Seminar, Cape Town, South Africa, March 31 – April 1, 2009.

[141] Wannek C, Kohnen B, Oetjen H.F, Lippert H, and Mergel J. (2008), Durability of ABPBI-based MEAs for High Temperature PEMFCs at Different Operating Conditions, Fuel Cells, 8: 87–95. doi: 10.1002/fuce.200700059.

[142] Litt M, Ameri R, Wang Y, Savinell R, Wainwright J, Proceedings of the Materials research society, 548, 313 (1999).

[143] PhD dissertation of Ameri, CWRU, Cleveland, USA, 1997.

[144] Okae I, Kato S, Seya A, Kamonoshita T, The Chemical Society of Japan 67th Spring Meeting, 1990, pp. 148.

[145] Seel D.C, Benicewicz B.C, Xiao L, Schmidt T.J, High temperature Polybenzimidazole based membranes, in Handbook of fuel cells, Vol.5, Fundamentals, Technology and applications, eds: Vielstich W, Gasteiger H, Yokokawa H, ISBN: 978-0-470-72311-1.

[146] Scholta J, Kuhn R, Wazlawik S, Jörissen L, Startup procedures for a HT PEMFC Stack, ECS transactions, 17 (1) 325-333 (2009).

[147] Ma Y, The fundamental studies of Polybenzimidazole/Phosphoric acid polymer electrolyte for fuel cells, PhD Thesis, RPI, New York, 2004. pp.125.

[148] Fontana B.J, The vapour pressure of water over phosphoric acids, J.Am.Chem.Soc., 1951, 73 (7), pp 3348-3350.

[149] MacDonald David.I, Boyack J.R, Density, electrical conductivity and vapour pressure of concentrated phosphoric acid, J.Chem.Eng. Data, 1969, 14 (3), pp 380 - 384.

[150] Brown Earl.H, Whitt Carlton D, Vapour pressure of phosphoric acids, Ind. Eng. Chem, 1952, 44 (3), pp 615 – 618.

[151] Schmidt T.J, PEMEAS, Frankfurt-Hoechst, Internal communications with ZBT Duisburg, in the course of collaborateve projects, 2005.

[152] Chin D.T, Chang H.H, On the conductivity of phosphoric acid electrolyte, Journal of applied electrochemistry 19 (1989) 95 – 99.

[153] Bandlamudi George, Jens Burfeind, Peter Beckhaus, Angelika Heinzel, Analysing electrolyte loss patterns while operating HT PEMFCs containing H_3PO_4 based electrolytes, European fuel cell forum, Lucern, Switzerland, July 2009.

[154] Kerres J, Schönberger F, Chromik A, Häring T, Li Q, Jensen J.O, Pan C, Noye P and Bjerrum N.J, Partially fluorinated arylene polyethers and their ternary blend membranes with PBI and H_3PO_4. Part I: Synthesis and Characterisation of Polymers and Binary blend membranes, Fuel cells (from fundamentals to systems) 3-4/2008, Wiley-VCH, weinheim.

[155] Higgins C.E, Baldwin W.H, Analytical Chemistry, 27, 1780 (1955).

[156] Vogel J, Development of Polybenzimidazole based high temperature membrane and electrode assemblies for stationary and automotive applications, DoE report, 2008.

[157] Oono Y, Fukuda T, Sounai A, Hori M, Influence of operating temperature on cell performance and endurance of a high temperature proton exchange membrane fuel cells, Journal of Power Sources, 195 (2010) 1007-1014.

[158] Okae I, Kato S, Seya A, Kamonoshita T, The Chemical Society of Japan 67th Spring Meeting, 1990, pp. 148..

[159] Kundler Isabel, Internal communications between ZBT Duisburg, Germany and BASF fuel cells, Germany, 2006.

[160] Nart F, Vielstich W, Handboook of Fuel Cells – Fundamentals, Technology and Applications, Vol.2, Electrocatalysis, Eds: Vielstich W, Gasteiger H.A, Lammy A, John Wiley & Sons.

[161] Hamann C.H, Vielstich W, Electrochemie, 4th Edition, Wiley-VCH, 2005, pp 650.

[162] Shi Y, Horky A, Polevaya O, Cross J, Book of Abstracts, Fuel Cell Seminar, 2005, Palm Springs, USA.

[163] Ferreira P.J, la O' G.J, Shao-Horn Y, Morgan D, Makharia R, Kocha S and Gasteiger H.A, Instability of Pt/C Electrocatalysts in Proton Exchange Membrane Fuel Cells. Journal of the Electrochemical Society 152 (11) (2005) pp. A2256-2271.

[164] de Bruijn F.A, Dam V.A.T, Janssen G.J.M, Review: Durability and degradation issues of PEM fuel cell components, Fuel Cells 08, 2008 (1): 3 – 22.

[165] Mehta V, Cooper J.S, Journal of Power Sources 114 (2003) 32 – 53.

[166] http://www.schunk-group.com/en/sgroup/BipolarPlates-FuelCells/schunk01.c.42453.en

[167] Heinzel A, Mahlendorf F, Niemzig O, Kreuz C, J. Power Sources 131 (2004) 35 – 40.

[168] Mueller A, Kauranen P, von Ganski A, Hell B, J. Power Sources 154 (2006) 467 – 471.

[169] Hung Y, Tawfik H, Mahajan D, J.Power Sources 186 (2009) 123 – 27.

[170] Joseph S, McClure J.C, Chianelli R, Pich P, Sebastian P.J, Int.J.Hydrogen Energy, 30 (2005) 1339 – 1344.

[171] Hung Y, El Khatib K.M, Tawfik H, J. Power Sources 163 (2006) 509 – 513.

[172] Kumara A, Reddy R.G, J. Power Sources 129 (2004) 62 – 67.

[173] Derieth T, Bandlamudi G, Beckhaus P, Kreuz C, Mahlendorf F, Heinzel A, Develop-ment of highly filled graphite compounds as bipolar plate materials for low and high temperature PEM fuel cells, J. New Mat. Electrochem. Systems, 11, 21-29, (2008).

[174] Heinzel A, Mahlendorf F, Beckhaus P, Kreuz C, Derieth T, Bandlamudi G, Burfeind J, Advances in materials for HT-PEM applications, Fuel cell seminar, Hawaii, USA, Novermber, 2006.

[175] Bandlamudi George, Derieth Thorsten, Mahlendorf Falko, Angelika Heinzel, Behav-ioural patterns of LT PEMFC and HT PEMFC over 1200 hours of operation, 10th Grove fuel cell symposium, London, 2007.

[176] Derieth T, Bandlamudi G, Beckhaus P, Heinzel A, Accelerated life time testing of com-pound based low and high temperature bipolar plates, Scientific Advances in Fuel Cell Systems Conference, Copenhagen 8th – 9th October, 2008.

[177] Stevens D.A, Hicks M.T, Haugen G.M and Dahn J.R, Ex Situ and In Situ Stability Studies of PEMFC Catalysts - Effect of Carbon Type and Humidification on Degrada-tion of the Carbon, Journal of the Electrochemical Society 152 (12) (2005) pp. A2309-A2315.

[178] Wang X, Kumar R and Myers D.J, Effect of Voltage on Platinum Dissolution - Rele-vance to Polymer Electrolyte Fuel Cells. Electrochemical and Solid-State Letters, 9(5) (2006) pp. A225-A227.

[179] Personal communications with BASF, Germany and Advent Technologies, Greece.

[180] Landsman D.A and Luczak F.J, Catalyst studies and coating technologies, Handbook of Fuel Cells - Fundamentals, Technology and Applications, W. Vielstich, Lamm, and H. Gasteiger, Editors, Vol. 4, pp. 811-831, John Wiley & Sons, U.K.   (2003).

[181] Aragane J, Murahashi T and Odaka T, Change of Pt Distribution in the Active Compo-nents of Phosphoric Acid Fuel Cell, Journal of the Electrochemical Society 135 (1988) pp. 844.

[182] Blurton K.F, Kunz H.R and Rutt D.R, Surface Area Loss of Platinum supported on

Graphite, Electrochimica Acta 23 (1978) p. 183.

[183] Gruver, G.A, Pascoe R.F, and Kunz H.R, Surface Area Loss of Platinum Supported on Carbon in Phosphoric Acid Electrolyte, Journal of the Electrochemical Society 127 (1980) pp. 1219.

[184] Honji A, Mori T, Tamura K and Hishinuma Y, Agglomeration of Platinum Particles Supported on Carbon in Phosphoric Acid. Journal of the Electrochemical society 135 (1988) pp.355.

[185] Bett J.A.S, Kinoshita K and Stonehart P, Crystallite Growth of Platinum Dispersed on Graphitized Carbon Black, Journal of Catalysis 35 (1974) pp.307.

[186] Mukerjee S and Srinivasan S, Enhanced electrocatalysis of oxygen reduction on platinum alloys in proton exchange membrane fuel cells, Journal of Electroanalytical Chemistry 357 (1993) pp. 201.

[187] Gasteiger H.A, Kocha S.S, Sompalli B and Wagner F.T, Activity benchmarks and requirements for Pt, Pt-alloy, and Non-Pt oxygen reduction catalysts for PEMFCs, Applied Catalysis B: Environmental 56 (1-2), (2005), pp. 9-35.

[188] Reiser C.A, Bregoli L, Patterson T.W, Yi J.S, Yang J.D, Perry M.L, Jarvi T.D, Electro chem. Solid – State Let. 2005 (8) 273.

[189] Borup R.L, Davey J.R, Garzon F.H, Wood D.L, Inbody M.A, J. Power Sources, 163 (2006), 76.

[190] Shao Y, Liu J, Wang Y, Lin Y, J. Mater. Chem. Volume 19 (2009), pp.46.

[191] Taniguchi A, Akita T, Yasuda K, Miyazaki Y, J. Power Sources, 130 (2004), 42.

[192] Knights S.D, Colbow K.M, St-Pierre J, Wilkinson D.P, J. Power Sources, 127 (2004), 127.

[193] Schmidt T.J, Durability and Degradation in High-Temperature Polymer Electrolyte Fuel Cells, ECS Transactions 1(8), 19-31(2006).

[194] Henschel C, Project meeting of the NextGenCell Project, Brussels, 24th March, 2009.

[195] Chhina H, Campbell S, Kessler O, Thermal and electrochemical stability of tungsten carbide catalyst supports, Journal of Power Sources, 164 (2), 2007, Pages 431-440.

[196] Santos L.G.R.A, Freitas K.S, Ticianelli E.A, Journal of Solid State Electrochemistry, Volume 11, Number 11, 1541-1548, 2007. DOI: 10.1007/s10008-007-0350-0.

[197] Yumura T, Kimura K, Kobayashi H, Tanaka R, Norio O.N, Yamabe T, The use of nanometer-sized hydrographene species for support material for fuel cell electrode catalysts: a theoretical proposal. Phys. Chem. Chem. Phys. 2009 (11) pp. 8275–84.